The Making of a Mining District

GREAT LAKES BOOKS
*A complete listing of the books in this series
can be found online at http://wsupress.wayne.edu.*

Philip P. Mason, Editor
Walter P. Reuther Library, Wayne State University

Dr. Charles K. Hyde, Associate Editor
Department of History, Wayne State University

*Title page frontispiece: The Copper Rock on the Ontonagon River, as drawn by
David B. Douglass, a member of the Cass expedition of 1820 and as engraved and
published in the American Journal of Science in 1821. For well over three centuries
the copper boulder has been the symbol of the copper riches of the Keweenaw.*

The Making of a Mining District

Keweenaw Native Copper
1500–1870

DAVID J. KRAUSE

 WAYNE STATE UNIVERSITY PRESS   DETROIT

Copyright © 1992 by Wayne State University Press,
Detroit, Michigan 48202. All rights are reserved.
No part of this book may be reproduced without formal permission.

08 07 06 05 04   6 5 4 3 2

Library of Congress Cataloging-in-Publication Data

Krause, David J.
   The making of a mining district : Keeweenaw native copper 1500–1870
/ David J. Krause.
      p. cm. — (Great Lakes books)
   Includes bibliographical references and index.
   ISBN 0-8143-2406-1. (alk. paper) — ISBN 0-8143-2407-X (pbk. alk.
paper)
   1. Keweenaw Peninsula (Mich.)—History. 2. Mining districts—Michigan—
—Keweenaw Peninsula—History. 3. Copper mines
and mining—Michigan—Keweenaw Peninsula—History. I. Title. II. Series.
F572.K43K73   1992
977.4′99—dc20   92-14640

Designer: Joanne E. Kinney

To DONALD F. ESCHMAN
Professor Emeritus
Department of Geological Sciences
The University of Michigan

CONTENTS

ILLUSTRATIONS	10
PREFACE	11
INTRODUCTION	14
1. KEWEENAW COPPER BEFORE 1800	18
Early Accounts	18
Champlain	24
The Jesuits	28
De la Ronde	34
The British Period	38
2. AMATEURS, PROFESSIONALS, AND NATIVE COPPER	44
The Creation of the Keweenaw	44
The Copper Lodes	45
Copper and Its Ores	48
Science—Amateurs and Professionals	54
3. HENRY ROWE SCHOOLCRAFT MEETS THE KEWEENAW	60
The Keweenaw Becomes American	60
Henry Rowe Schoolcraft	68
The Expedition of 1820	71

The Schoolcraft Reports	77
American Responses	90

4. Douglass Houghton—Copper Finds Its Columbus 97
Early Life and Education	97
The Expedition of 1831	103
The Expedition of 1832	108
The Quiet Years	111
The State Geologist	117

5. The Copper Report and the Copper Rush 122
The 1840 Field Season	124
The Copper Report	132
The Boulder and the Rush	136
The Fateful Journey	143

6. Houghton—The Misunderstood Pioneer 149
The Houghton Tradition	152
Questioning the Tradition	156
Testing the Tradition	160
Houghton—An Alternative Perspective	171

7. Charles T. Jackson and the Federal Survey 176
Charles T. Jackson	182
The Priority Conflicts	184
The Federal Survey Fiasco	187

8. Jackson and the Early Mining Efforts 194
The Jackson Claim	194
The Mining Efforts Begin	205
The Cliff Strikes It Rich	209
The Mighty Minesota	214
The Foster and Whitney Report	219

9. THE DISTRICT BECOMES ESTABLISHED	221
The Spectacular Fissures	221
The Amygdaloids Produce	229
The Conglomerate Bonanza	234
10. THE MAKING OF A MINING DISTRICT	239
NOTES	252
REFERENCES CITED	275
INDEX	285

ILLUSTRATIONS

    The copper rock on the Ontonagon River   Frontispiece
1. *The Great Lakes and St. Lawrence River regions*   / 21
2. *The Lake Superior region*   / 27
3. *The Jesuit map of Lake Superior of 1672*   / 33
4. *The Carver map of 1778*   / 42
5. *The geology of the Keweenaw peninsula*   / 46
6. *The three copper lodes of the Keweenaw*   / 47
7. *Large mass of native copper from Isle Royale*   / 53
8. *Lewis Cass*   / 66
9. *Henry Rowe Schoolcraft*   / 69
10. *Route of the Cass Expedition of 1820*   / 72
11. *The Cass expedition at the Ontonagon Boulder*   / 75
12. *Douglass Houghton*   / 98
13. *Houghton's sketch of the green rock location*   / 126
14. *Houghton's field notes describing the Ontonagon Boulder*   / 128
15. *Houghton's map and section of the Portage Lake region*   / 131
16. *The Ontonagon Boulder at the Smithsonian Institution*   / 139
17. *Charles T. Jackson*   / 183
18. *Early mining locations on the Keweenaw peninsula*   / 200
19. *The Lake Superior Company copper vein on Eagle River*   / 203
20. *The Lake Superior Company stamp mill*   / 204
21. *John Hays*   / 208
22. *The Cliff Mine*   / 213
23. *Early sketch of the Minesota vein*   / 215
24. *Section of the Minesota Mine in 1849*   / 217
25. *The Minesota Mine in the 1850s*   / 218
26. *Miner and mass copper underground*   / 223
27. *Ransom Shelden*   / 231
28. *Edwin J. Hulbert*   / 235
29. *General view of Calumet*   / 241
30. *Michigan and U. S. copper production*   / 242
31. *The Quincy location in the 1960s*   / 245

PREFACE

This book is the outgrowth of a long-standing interest in the Keweenaw copper district that began with my first visits to the copper country as a teenager in the late 1950s. After then reading Angus Murdoch's *Boom Copper* I became intrigued by the history of the area, by the unusual nature of the native copper deposits, and by stories of the exploits of Douglass Houghton, Michigan's first state geologist. The opportunity to expand that interest into a more formal study, however, came only many years later when members of the Department of Geological Sciences at The University of Michigan indicated that a historical study could form the basis of a doctoral dissertation. At first I concentrated on examining Houghton's geological efforts in the copper district. This led to a closer consideration of the endeavors of Henry Rowe Schoolcraft and Charles T. Jackson. From there the study grew into an examination of how the gradual recognition of the unique nature of the Keweenaw native copper deposits was accomplished between 1500 and 1870, from the earliest contacts of Europeans with the New World to the full establishment of the region as a major copper-producing district.

When that work was completed in 1986 several persons suggested that the results of my research should be made available in a more generally accessible form. This book is the result, and is based on my dissertation but differs from it in a variety of ways. In the interests of readability, wording changes and adaptations of language have been made, several sections have been revised or deleted, many photographs and illustrations have been added, and the notes have been rearranged. However, all significant historical conclusions are included and are fully documented in the notes.

The role of the Keweenaw copper district in the exploration and development of the New World and of the United States deserves more attention than it has been given so far. I have therefore tried to view that role from a broadly based perspective, emphasizing the importance of the district in the opening of the American interior and its impact on the national consciousness. Also, the first half of the nineteenth century was a time of transition in American science from an amateur to a professional perspective, and I have examined the effects of this change on attempts to come to grips with the nature of the copper deposits of the Keweenaw. My studies have led me to conclude that certain aspects of the early history of the copper district need revision, and that the role of Douglass Houghton, one of the most revered figures in that history, has in certain particulars been significantly overestimated. I hope that these issues will be of interest not only to those of an academic or scientific background but also to those whose primary interest is simply the history of this fascinating region.

Several persons deserve acknowledgment for their part in making this book possible. I thank Prof. Rob Van der Voo, chairman of the Department of Geological Sciences at The University of Michigan, for his support of a historically based dissertation topic, and Prof. Nicholas H. Steneck of the Department of History for his advice and direction during the original study. Prof. Donald F. Eschman, also of the Department of Geological Sciences, was a cochairman of my dissertation committee, and for many years he encouraged my geological interests. I express special appreciation to him by the dedication of this book. Dale Austin helped greatly in making the various diagrams legible. Finally, I thank the staff members of the Wayne State University Press for their invaluable guidance in seeing the manuscript into print.

I alone am responsible for any errors of fact or interpretation that remain. I would be pleased if the views I have expressed here about events and people would encourage others to consider alternative perspectives. A book that includes historical analysis is never the final word on any subject. Hopefully it is instead one further step in the direction that leads toward a more complete understanding of the past, and therefore also of the present.

INTRODUCTION

Lake Superior is the largest body of fresh water on earth. It is also the highest and most remote of the five Great Lakes that provide an avenue into the heart of the North American continent. A long narrow peninsula extends northeastward from the southern shore of the lake toward its center. In earlier times, as the Indians passed along that shore in canoes, the peninsula acted as a barrier to their travels and forced them outward into a long and possibly dangerous detour. Near the center of the peninsula, however, there was a river and lake passage between its southeastern and northwestern coasts that could be negotiated by small craft. Because, aided by only a short portage, many miles could thereby be cut from the journey, they named the region "Kakiweonan," meaning "the place where they traverse a point of land" or possibly "the place of the detour."[1] After repeated transliterations the name finally became Keweenaw (pronounced "KEE wa naw"), but to the Indians the peninsula was more than just an obstacle in their path. For centuries they had also been aware of the pure copper that could be found along its shores and in the adjacent regions around the western end of the great lake.

With the passing of the years and the coming of the Europeans from across the sea, this copper increasingly attracted attention, and encouraged many to speculate about the potential mineral wealth of the interior of the American continent.

Introduction

These speculations culminated in the 1840s when the Keweenaw, by this time a part of the state of Michigan's northern, or "upper," peninsula, became the focus of North America's first great mining boom. The opening of that district abruptly transformed the United States from a nation vexingly dependent on foreign imports of copper into the world's greatest producer of the red metal, and contributed significantly to the emergence of America as an industrial and technological power. Although production virtually ceased some decades ago, the district remained a significant source of copper for more than a century, producing well over ten billion pounds. A historian of mining has put it succinctly: "It is quite possible that the Keweenaw was in the long run the most profitable of any American nonferrous mining district."[2]

And yet, over the years, this "most profitable" district has not received from historians the attention that its status would seem to merit. Several possible reasons for this neglect can be suggested. In the minds of many the California gold rush of 1849, which followed the Lake Superior copper rush by only half a decade, has become *the* mining rush of nineteenth century America, and it must be granted that the glitter of gold is not easily matched by a metal such as copper. Others might argue that the Lake Superior region was just a sort of "sideshow" in the westward drive to the Pacific which was such a dominant part of the national consciousness in the last century. However, a curious additional reason for slighting the history of the Keweenaw has been offered by the same historian quoted above as acknowledging the district's economic importance: "Perhaps the reason for its lack of fame outside Michigan itself is that its annals, like those of happy peoples, are relatively dull."

Perhaps this is so, but before the judgment of "dull" with its apparent justification of neglect can be accepted, care should first be taken to ensure that those annals have been fairly examined and properly understood. I will therefore try to provide here, from the perspective of one whose background is in geology rather than history, an accounting of the earlier "annals" of this region, at which time people were looking forward with uncertainty or anticipation to that time when the Keweenaw would come into its own as a significant copper-producing district. The later history of the Lake Superior copper-mining industry, from its beginnings about 1850 to its demise shortly before 1970, has now been very well analyzed by Larry Lankton in his *Cradle to Grave: Life, Work, and Death at the Lake Superior Copper Mines*.³ Not the least of my hopes is, as it was also of Professor Lankton, that works such as these will help ensure that future judgments on the significance of the Keweenaw will be based on a more accurate and complete understanding of its history.

In this book I will examine the Keweenaw copper district from the time that Europeans first became aware of its existence shortly after the discovery of the New World by Columbus to the full establishment of the district as one of the world's great copper producers about 1870. Two major themes will provide threads of continuity that can be followed through the fabric of people and events that spans this interval of nearly four centuries.

The first theme is the uniqueness of the native copper found in the Keweenaw. In all other copper-mining districts of the world, past or present, most of the copper is found chemically combined with other elements. When taken from the earth, the copper-bearing minerals found in these other districts do not in any way resemble metallic copper. They must first be subjected to complex smelting processes that free the copper from the other elements with which it was combined. However, when mines were opened on Lake Superior, it was discovered that all the copper was already in the pure metallic "native" state, having been refined by nature itself.⁴ This unexpected and unprecedented characteristic made the Keweenaw unique in the history of copper mining in a fundamental way, but as we shall see, recognizing this apparently simple fact was not easy for many persons. The path toward the profitable mining of native copper had many twists and turns, and was tangled as much by what people *expected* to find as by what in fact was there. A new way of looking at the world was needed, and at least for some the conceptual revisions were never achieved.

The second thread of continuity is the contrasting approach that two groups, the "amateurs" and the "professionals," used in trying to grasp and decipher this uniqueness. The half-century that preceded the establishment of the Lake Superior mining industry was also the time when science in America was shifting from being a pleasant diversion for amateurs into a more rigorous domain of the professionals, with their presumed greater insights into the workings of nature. And yet the amateurs who knew the least about how copper was normally found in a mining region were most open to the unforeseen possibilities that Keweenaw copper presented, while most of the new professional geologists, despite their greater training and background, could not see any value in this unusual form of copper until that was actually demonstrated by successful mining efforts.

These themes will be developed most fully by an examination of the efforts of three persons who investigated the Keweenaw during the critical interval from 1820 to 1850: Henry Rowe Schoolcraft, explorer of the upper Great Lakes and Mississippi valley regions; Douglass Houghton, Michigan's first state geologist; and Charles T. Jackson, a geologist who explored the district near midcentury. Although the efforts of each of these men in the Keweenaw have been examined in earlier studies, several authors have made claims about their work that do not stand up well when the evidence is examined.[5] In particular, and despite the reputation he has achieved through the years, I will argue that a revision is called for of the commonly accepted story of Douglass Houghton's role in understanding the copper deposits on Lake Superior. Such a revision will also throw a new and different light on the pioneering efforts of both Schoolcraft and Jackson.

In analyzing these persons and events, however, there would be little of value in a study that simply applauded those who foresaw the "right" answers to difficult questions about nature and denounced those whose ideas were "wrong" if judged by the standards of today. The past should be approached on its own terms as far as possible. We should never forget that what may be clear or obvious to us now was not always so, and that we today view the past with the incomparable luxury of hindsight. Nevertheless, the Keweenaw district and its abundance of native copper was and is unique on this earth. This uniqueness can therefore be a firm background against which the events of the past can be projected and viewed, helping us to illuminate the history of a most interesting and noteworthy region of the natural world.

1

KEWEENAW COPPER BEFORE 1800

Five hundred years ago the times were changing. The printing press had helped create a new interest in learning about the natural world, and the excitement of discovery was increasingly in the air. Sailing across the seas in search of an old world, Columbus had stumbled on a new one but, ironically, he could not accept that fact. He died believing that he had reached the East, never understanding what it was that he had found, and it was left for his successors to define, exploit, and settle the land that he had only glimpsed. To the south, the Spanish found this New World to be immensely rich in gold, silver, and pearls, and naturally the tales grew as they were repeated. To the north, where the French and the English were to find their foothold, the land was more somber. It was heavily forested and the winters were severe, but rumor had it that this land too was rich beyond belief, and why not? Immense diamonds were there, said the stories. There was gold too, of course, and also, it was noted very early, there was copper.

Early Accounts

The copper tales spoke, vaguely to be sure, of a land on the shores of a remote freshwater sea, far toward the interior of the North American continent, where the metal lay scattered over the ground in huge pure nuggets, already refined by nature. Copper, however, stood little chance against the urge for gold and diamonds. However, as the years passed most of the romantic rumors were reluctantly but inevitably replaced by the more prosaic facts. The diamonds turned out to be quartz, and the gold was only "fool's gold," worthless iron pyrite. But the copper? Amazingly, the copper was there, just as the rumors proclaimed.

The northern regions of the New World along the Atlantic coast had been explored long before by the Norsemen, who had even attempted to establish a colony, but those preliminary efforts had never become a part of the general geographic knowledge of Europe. At the time of Columbus, however, fishermen already knew of, and were exploiting for the tables of Europe, the rich harvests of fish on the northern shelves off the American continent, but the lands behind the shores remained mostly unknown. The possibility of a passage through this strange landmass to the fabled treasures of the Far East, however, proved a powerful lure. Slowly, the contours of the new continent began to emerge, even though it was at first regarded only as a barrier to the real goal that lay beyond.

Columbus had chosen to seek the riches of the East by sailing westward from Spain. One of the first to try a more northerly route to Asia was another Genoese, John Cabot, who sailed from England with his son Sebastian. They coasted and made a landing in America near Newfoundland in June 1497, and their return to England caused great excitement. The future prospects looked bright, and in 1509 Sebastian Cabot made his second voyage to America, following the Atlantic coast as far southwestward as the present Cape Hatteras.

Subsequent probing voyages by the English, however, revealed little that resembled the riches or civilization that they expected of Asia. There was no silk, gold, or spices, and the cultural level of the natives left little promise of trade or wealth. Spain, it seemed, had been particularly fortunate in the south. "The equator is rich," but apparently the North was not, and the English efforts in the westward direction lost their impetus for many years. Even so, in an account of Cabot's 1509 voyage written a few years later, there is a casual, passing observation of interest here: "Many say that they have seen copper in places in the hands of the inhabitants."[1]

References to copper possessed by the natives became a theme, although admittedly one in a minor key, in the accounts of many explorers over the next three centuries. Although copper was the third, after gold and silver, in the "triumvirate of precious metals" sought by the early explorers of the New World, it lacked the glamour of the other two. Although it was an essential resource of the day, references to copper are found frequently but unsystematically in the accounts of those whose real interests lay largely in other directions, and who would much rather have found large quantities of the other, more valuable metals.

The French became part of the great westward movement in late 1523, and Francis I, the French king who could not help envying the wealth that the Spanish were finding in the South, sent out Giovanni da Verrazzano to see if he could find the hoped-for northwest passage to Asia. Verrazzano's careful probing of the Atlantic coast of America is best known for his entrance into what is now the harbor of New York City. In his descriptions of the native peoples he encountered, he noted that some had beads of copper in their ears while others had "sheets of worked copper which they prize more than gold," and he speculated that hills in the region might contain the metal.[2] In 1525 Estavo Gomes, a Portuguese pilot sailing in the service of Spain, made another attempt at the elusive passage through the continent. He too was unsuccessful, but on his return he also noted that the Indians said that they had silver and copper.[3]

Where did this copper come from? A variety of copper ores is found at many places up and down the Atlantic seaboard, but it seems unlikely that the Indians would have had the technological skills to smelt the copper from those ores. In some places small amounts of pure native copper are also found, and it possibly may have been used by those original Americans. There is no simple way, at this late date, to determine exactly where the copper seen so early in the hands of the Indians had its origin, although the argument can be made that all of it came from the Keweenaw.[4] To this point, however, only the fringes of the new land had been touched. The Keweenaw copper district and the huge "superior" lake whose waves lapped its shores were still part of the vast unknown wilderness beyond. The northern Atlantic coastline was a maze of estuaries, bays, and islands, and the discovery of the great river that would become a road into the interior was necessary before the mystery of what lay beyond could be unraveled.

The St. Lawrence River is clearly the primary water route into the Great Lakes and the heart of the North American con-

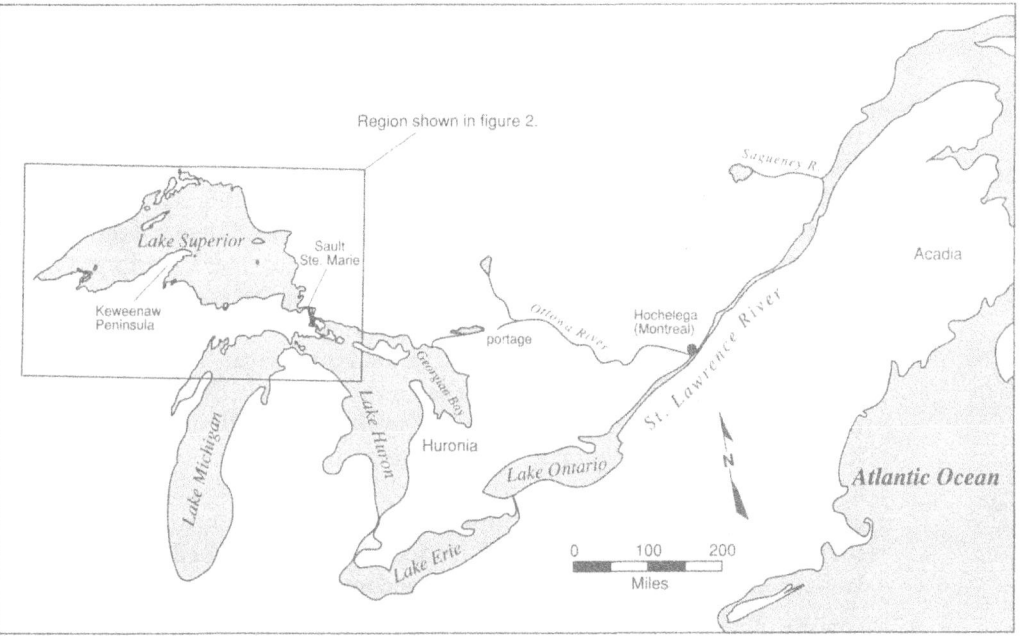

Fig. 1. *The Great Lakes and St. Lawrence River regions.*

tinent (fig. 1). The river, however, flows into an ever widening bay that reaches the ocean in a confusion of islands, inlets, and peninsulas. Early navigators certainly passed through this bay, the Gulf of St. Lawrence, but finding the actual entrance to the river proved more elusive. Jacques Cartier, a prominent citizen of the French seaport of Saint Malo, carried the flag of France a thousand miles into the new continent as he searched for the missing passage to the East.

Francis I again provided the needed support, as his hope of finding gold and other wealth continued unabated. Cartier sailed in 1534 and found temptingly fertile lands along the southern part of the gulf, but in the confusion of points, bays, and the frequent fogs he missed the river. Claiming the lands for France, he returned home with little of substance to report. Two Indians who returned with him to France, however, told him about the river and, far upstream, the kingdom of Hochelaga. They also spoke of gold and copper in the "country of the Saguenay" and they offered to take Cartier there on a return trip. Once more the king was filled with the thoughts of riches.

The following year Cartier again crossed the Atlantic, this time sailing far into the gulf. As he proceeded westward toward the entrance to the river, the Indians told him "that this was the beginning of the Saguenay and of the inhabited region; and that thence came the copper they call *caignetdazé*."[5] On the first of September they came to the mouth of "a very deep and rapid river, which is the river and route to the kingdom and country of the Saguenay." Proceeding southwestward up the river, whose shores were at the peak of their fertile lushness, they finally reached the village of Hochelaga. There the Indians took him to the top of the mountain that overlooked the village. Cartier named it Mont Royal (Montreal), and from its height he gazed over a wide expanse of country with the great river stretching even farther to the southwest as the Indians told to him what they knew of the region. Later, after returning back down the St. Lawrence, Cartier watched as an Indian chief, who had agreed to accompany them back to France, was presented with going-away gifts by his subjects, one of which was "a large copper knife from the Saguenay."[6] An account of what he heard illustrates well the difficulties we now have in understanding exactly what the exchange was all about, even if we are looking at a modern map of the region (fig. 1). They told him about this "Saguenay," a word that was apparently being used in two different senses. They said:

> that the above-mentioned river, named the "river of the Saguenay", reaches to the (kingdom of the) Saguenay, which lies more than a moon's journey from its mouth, towards the west-north-west; but that after eight or nine days' journey, this river is only navigable for small boats; that the regular and direct route to the (kingdom of the) Saguenay and the safer one, is by the river (St. Lawrence) to a point above Hochelaga where there is a tributary [clearly the Ottawa River], which flows down from the (kingdom of the) Saguenay and enters this river (St. Lawrence) as we ourselves saw, from which point the journey takes one moon. And they gave us to understand, that in that country, the natives go clothed and dressed in woolens like ourselves; that there are many towns and tribes composed of honest folk who possess great stores of gold and copper. Furthermore they told us that the whole region from the first-mentioned river up as far as Hochelaga and (the kingdom of the) Saguenay is an island, which is encircled and surrounded by rivers and by the said (St. Lawrence); and

that beyond the (kingdom of the) Saguenay, this tributary [the Ottawa] flows through two or three large, very broad lakes, until one reaches a fresh-water sea, of which there is no mention of anyone having seen the bounds, as the people of the (kingdom of the) Saguenay had informed them; for they themselves, they told us, had never been there.[7]

This account is both informative and confusing. On the one hand "Saguenay" was, and still is, the name of a river that entered the St. Lawrence from the north over 250 miles downstream, back northeastward toward the Atlantic Ocean. On the other hand, Saguenay was apparently also being used as the name of a kingdom, referring to lands to the west, in the direction of the lake now called Huron. The reference to the Ottawa River, which enters the St. Lawrence from the west just above Montreal, seems explicit. This river follows a former ice-age drainage channel that provided, with portages, access to Georgian Bay on Lake Huron and, ultimately, to the shores of Lake Superior far beyond. This route bypassed completely Lakes Ontario and Erie, which lay as yet unknown farther to the southwest. The "fresh-water sea" spoken of seems to be a clear reference to Lake Huron, and therefore this must be one of the very earliest references by any European to one of the Great Lakes.

The Indians were apparently not consistent in describing exactly where the copper had come from, but at least part of the confusion may have resulted from difficulties Cartier had in understanding what the Indians were saying. The references to "woolens" and "honest folk" seem garbled, and some of the supposed allusions to gold may well have been only in the minds of the Frenchmen.[8] Nevertheless, whatever the exact meaning was, a shadowy land to the west from which the copper had originally come seems to loom behind the stories.

After spending a severe winter in the new land, Cartier returned to France in the spring of 1536, bearing the exciting news of gold, silver, copper, and the annexation of a vast country. In 1541 another expedition west had the goal of settlement, but Indian hostility was becoming an increasing concern. The community was elated, however, when large quantities of both "gold" and "diamonds" were discovered at the very locality at which they had chosen to settle on the St. Lawrence. The gold was tested in the furnace and found to be of good quality, and the treasures were carefully packed into casks to await the return journey.

The cold winters and the Indian opposition took their toll, but the final blow came when the minerals reached France and the excited Francis I had them put to the official test. The gold, incredibly, turned out to be iron pyrite, "fool's gold," and the diamonds were simply crystals of quartz. "False as a Canadian diamond" became the mocking watchword, and the hopes of France for a colony rivaling those of Spain were crushed as the Spanish reiterated the theme that there was nothing of value in the North. For over a half-century French interest in the New World went into eclipse, and contacts were limited to those of a few fur-traders and fishermen.[9]

Champlain

The hiatus was finally ended by the efforts of Samuel de Champlain, the founder of Quebec and "the father of New France." Between 1603 and 1633 he established a firm French presence in the New World through a series of voyages to the St. Lawrence region. In 1603 he sailed up the great river to beyond Hochelaga at Mont Royal, but he found that the village seen by Cartier was gone. The Indians that Cartier had visited there apparently had been the victims of more warlike tribes that were engaged in the growing pursuit of furs. Champlain, as part of his long-range planning, set out to formulate an accurate, consistent geographical picture of the region based largely on what he learned from the Indians. From what he heard he was able to infer the existence of Hudson Bay seven years before it was found by Hudson.[10] He also heard of Lake Ontario, above the rapids of the St. Lawrence, and of Niagara Falls even farther beyond.

Minerals were still a major objective, and whenever he had the chance he asked the Indians about the location of mines. The Algonquins showed him some copper bracelets that they had obtained from a tribe called the "good Iroquois," who knew of "a mine of pure copper" somewhere in the North.[11] They seemed to indicate, however, that the mine was on a salt sea a great distance away, and Champlain chose not to take advantage of their offer to guide him to the place. Instead, for the next five years his efforts were concentrated in Acadia, the region along the southern bank of the St. Lawrence and the adjacent Atlantic coast, and the interior was temporarily set aside.

From the standpoint of Keweenaw copper these years were largely diversionary. Another explorer, Sieur de Prévert of St. Malo, told Champlain about Indian tales of mines of copper and other metals in the region of the Bay of Fundy, and any minerals so conveniently located naturally had to be thoroughly investigated. According to the stories these mines were not ordinary; the Indians claimed that they were guarded by "a dreadful monster named Gougou," who was twice as high as the ship's mast, and even Prvert himself claimed to have heard it hissing as he passed its den. Other mineral localities were also said to be not far, and the next year they searched for them again but had little luck, although they did find "two copper mines, not in the native state, but apparently that metal, according to the miner, who considered them very good." Later, they again looked for the mine of pure copper, but "when we reached the spot where we hoped was the mine we were seeking, the Indian could not find it."[12] The mines of Acadia proved to be very elusive.

In 1608 Champlain was back on the St. Lawrence, being increasingly drawn into the conflicts between the Indian tribes that were a result of the expanding fur trade. In 1610 the Indians he was supporting "promised to show me their country and the great lake, and some copper mines." That June, two Indians met him on the St. Lawrence. His remarks on the meeting are of interest:

> After conversing with them a short time about a number of things touching their wars, the Algonquin Indian, who was one of their chiefs, drew out of a sack a piece of copper a foot long, which he presented to me. It was very fine and pure. He gave me to understand that the metal was abundant where he had obtained it, which was on the bank of a river near a large lake. He said that it was taken out in pieces, and when melted was made into sheets and smoothed out with stones.[13]

This was undoubtedly Lake Superior copper, although Champlain may have misunderstood the process they were describing. It is unlikely that any working of the copper by the Indians involved actual melting, although some forging of the heated metal may have been done.[14] More intriguing is the description of the location from which the copper had come: the bank of a river near a large lake, where the metal was abundant. It is impossible to conclude anything certain about the exact location being described, but it is just barely possible that this is the first known reference to the great copper rock that lay on the bank of the Ontonagon River near the western end of Lake Superior. This huge mass of pure copper, the "Ontonagon Boulder," became a major factor in attracting later attention to the region, and it was to exert an almost mystical influence on the minds of many men for many years. "I was much pleased with this gift," he wrote, "although it was of small value."

One of Champlain's more significant accomplishments was sending French youths to live with the Indians and act as interpreters. This set up the close association with the natives that was to be a distinguishing characteristic of the French period in the New World. One of those young men was Etienne Brulé, who had arrived in Quebec as a teen in 1608. He became the prototype of the French voyageurs and *coureurs de bois* who were to be so important in the following years. Champlain placed him in the care of the Algonquins, and until he was killed and eaten by the Indians in 1632 Brulé traveled extensively in the interior of North America. He also became, in all probability, the first European to see Lake Superior.

In 1615 Champlain finally found time to press westward again. Accompanied by Brulé and four Recollect monks, who were beginning the long association of Catholic priests with the Indians of the lakes, they finally reached the Sweetwater Sea, Lake Huron, which Champlain had so long sought. Brulé continued his own undocumented wanderings, and in 1623 Gabriel Sagard, a lay brother who became the historian of the Recollect efforts, reported what he had learned as a result of a journey that Brulé and another French youth named Grenoble had made:

> above Mer Douce [Lake Huron] is another very great lake, discharging into it by rapids nearly two leagues broad named Sault de Gaston. This lake with the Mer Douce has an extent of thirty days' journey by canoe, according to the report of the savages, and according to that of the interpreter (Brulé?) it is four hundred leagues in extent.[15]

This is a clear and explicit reference to the greatest of the Great Lakes, the upper, or "superior," lake (fig. 2). It seems very likely that Brulé had visited Lake Superior sometime before 1623, and that he may have named the rapids at the outlet of the lake, the present Sault Ste. Marie, after the brother of the King of France.

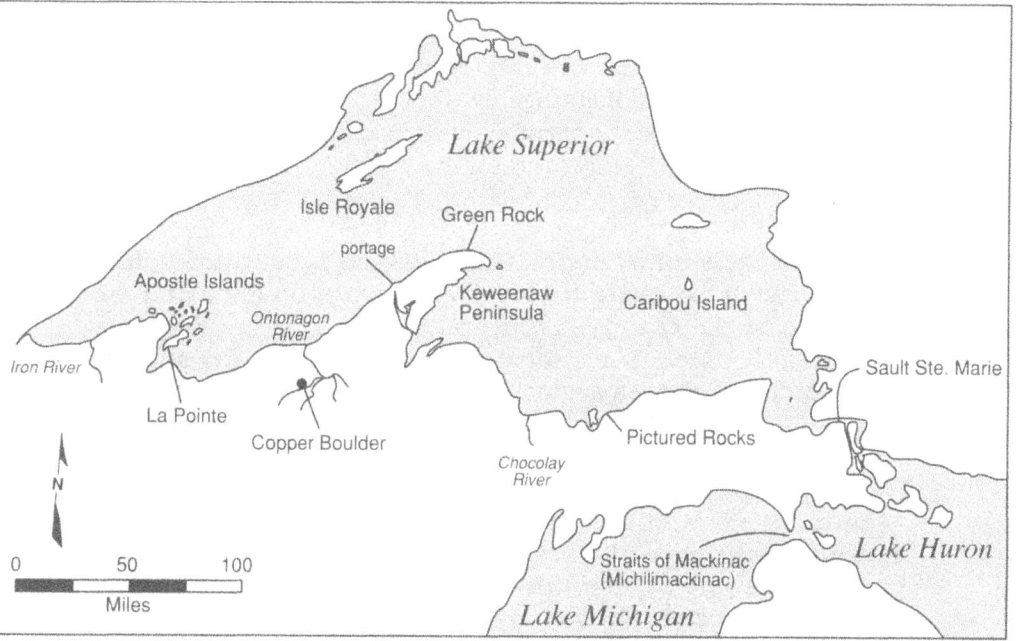

Fig. 2. The Lake Superior region.

Equally notable are Sagard's comments about the copper. Brulé and Grenoble evidently had been "trading among a nation towards the north, about a hundred leagues from us, getting copper from a mine."[16] Later, back in Paris, Sagard described the resources of the region:

> There are in addition copper mines not to be despised, from which profit might be derived if there were a population and workmen who were willing to work them conscientiously. This might happen if colonies had been established, for about eighty or a hundred leagues from the Hurons there is a mine of copper, an ingot from which was shown me by the interpreter [Brulé] on his return from a journey made in the district."[17]

It is difficult to know how seriously to take the distances cited by Sagard, but it seems at least possible that Brulé and Grenoble had entered Lake Superior and traveled extensively westward, perhaps even to the copper mass on the Ontonagon River or to Isle Royale, where the copper had been mined by the Indians in earlier times. They may well have been the first

Europeans not only to see Lake Superior but to see the copper of the Lake Superior region in the land of its origin. A major step had clearly been taken, although these Frenchmen of the wilderness were far more interested in occasions of the moment than in any long-range, systematic attempt to carry out Champlain's goals of a thriving New France.

The Jesuits

After this initial notice, the copper on Lake Superior does not seem to have attracted much attention again until after the middle of the century. Thereafter, however, it became more frequently noted. The efforts of the Recollects to Christianize the Indians had been augmented by the presence of the Jesuits, and they established a minor seat of culture in Huronia, the country of the Hurons near the southern shores of Georgian Bay on Lake Huron. The Jesuit *Relations*, the official records of their efforts in the New World published in Paris, abound with observations of all aspects of Indian life and the natural world. As part of their missionary efforts, the Jesuits became active explorers and traveled long distances toward the interior. In so doing they made many passing but important observations on the evidences of the copper that they encountered.

One of these Jesuits was Francesco Bressani, a missionary from 1645 to 1649. He returned to Italy after the attacks of the Iroquois, whose wars resulted in the destruction of the Hurons and the collapse of the mission. There he wrote an account of his experiences that included a description of the natural features of the country. In telling of their missionary efforts, he sketched a revealing account of the emerging picture of the region where Lakes Huron, Michigan, and Superior come together. His account also revealed his interest in the natural resources of the land:

> They are at least as many as nine [Indian nations],—one of them being the nation of the Sault, or cascade [the Sault Ste. Marie rapids], more than 300 miles distant from us, through which we hoped for a passage in order to reach other nations farther on, who dwell along a lake [Lake Superior] larger than the fresh-water sea [Lake Huron], which takes its origin thence, and extends between the West and the north. A Peninsula, or Strip of land [the upper peninsula of Michigan], divides this lake [Lake Superior] from the one which is called "lake of the Stinkards" [Lake Michigan]. . . . The earth contains iron ores, and certain

rocks which melt like metal, with an appearance of having some vein of silver. There is a Copper ore, which is very pure, and which has no need of passing through the fire; but it is in places far distant and hard to reach, which render its transportation almost impossible. We have seen it in the hands of the Barbarians, but no one has visited the place.[18]

Clearly, he was impressed with the native state of the copper and the fact that no smelting was needed. Its exact location, however, was to him still only vaguely known, and it appears even Brulé's visit was not widely regarded at the time.

In the summer of 1660 another Jesuit, apparently Father Druillettes, met an Indian named Awatanik, who had been one of his converts from a decade earlier. After the dispersion of the Hurons that followed their conquest by the Iroquois, the Indian had wandered for two years through the country around and north of Lake Superior.[19] He related his experiences to the father, who passed the information on to be included in their accounts that were published in Paris. "This lake . . . is also enriched in its entire circumference with mines of lead in a nearly pure state; with copper of such excellence that pieces as large as one's fist are found, all refined." The father's main interest, however, was in the sea the Indian reported to the south, west, and north. He concluded that "it is a strong argument and a very certain indication that these three Seas, being thus contiguous, form in reality but one Sea, which is that of China."[20] The ancient lure of the Far East still continued its hold on the minds of the time.

The attractions of the rich fur trade were drawing an ever increasing number of explorers and traders into the upper lakes region, among whom were Radisson and Groseilliers. In the fall of 1659 they paddled the length of the south shore of Lake Superior and their account added much to the knowledge of its geography. They took the Indian portage route across the middle of the Keweenaw peninsula, thereby avoiding the long journey around the point. They did note that "in the end of that point, that goeth very farre, there is an isle, as I was told, all of copper. This I have not seene [sic]." They also reported that the Indians were making the journey to "a great island far out into the lake," which was clearly Isle Royale.[21]

By using the reports of the traders, in 1664 Pierre Boucher described, in his history of Canada, what he had learned about the location of the copper in greater detail:

> In Lake Superior there is a large island [Isle Royale], about fifty leagues around, in which there is a fine mine of red copper; there are also found in several places large pieces of this metal in a pure state. . . . They have told me that they saw an ingot of pure Copper, weighing according to their estimate more than eight hundred pounds, which lay along the shore; they say that the savages when passing by, make a fire above this, after which they cut off pieces with their axes; one of the traders tried to do the same, and broke his hatchet in pieces.[22]

This "ingot" may well have been a large mass of copper that lay on the shore of the lake close to its western end, near the Iron River in what is now Wisconsin. It may have been even larger than the more famous mass on the bank of the Ontonagon River, but it did not survive into the nineteenth century.[23]

The Jesuits continued to be a conduit for information passing back to the Old World. About 1670, the Fathers Allouez and Dablon left accounts that further localized where the copper was found. Allouez's account appeared in the *Relations* published in Paris in 1668. Having traveled extensively among the Indians, he reported that "one often finds at the bottom of the water pieces of pure copper, of ten and twenty livres [pounds] weight. I have several times seen such pieces in the Savages' hands; and since they are superstitious, they keep them as so many divinities. . . ." He also spoke of the great rock of copper along the shore from which those who passed would cut fragments and of the "true red copper" from Isle Royale.[24]

Father Dablon amplified on this considerably, although he had not visited many of the places he described. He specifically declared that he did not necessarily believe everything that he had heard, and he was not above resorting to "artifice" to get some of the information that he wanted from the Indians. He described an imaginary tour around Lake Superior in a counterclockwise direction, pointing out where the metal was found. He told in some detail the Indian stories about the copper from Missipicouatong [Michipicoten] Island at the eastern end of Superior, "but farther toward the West, on the same North side, is found the island which is most famous for Copper, and is called Minong [Isle Royale]; this is the one in which, as the Savages have told many people, the metal exists in abundance, and in many places." There, the deposits are found "principally in a certain inlet that is near the end facing the Northeast. . . ." Continuing on to the end of the lake, and coming back a day's journey along the South side, [probably at the Iron River in Wisconsin] "one sees at the water's edge a Rock of Copper weighing fully seven or eight hundred livres. . . .'

After referring to the many masses of copper found on the surface in the Apostle Islands and to a "slab of pure Copper, two feet square, and weighing more than a hundred livers" that they bought from the Indians, he then went on to offer a theory for how these masses had come to be scattered across the region:

> It is not thought, however, that the mines are found in the [Apostle] Islands, but that all these Copper pebbles probably come from Minong [Isle Royale] or from the other Islands which are the sources of it, borne upon floating ice or rolled along in the depths of the water by the very impetuous winds, particularly by the Northeast wind, which is extremely violent.

Resuming his tour of the lake, he stated that on continuing eastward along the southern shore

> one enters the River called Nantounagan [Ontonagon], in which is seen a height from which stones of red Copper fall into the water or on the ground, and are very easily found. Three years ago we were given a massive piece of it, a hundred livers in weight, which was taken in this same spot; from it we have cut off some fragments, and sent them to Quebec to Monsieur Talon [the French intendant].

He spoke also of their desire "to examine a certain verdigris which is said to run down through the crevices of certain Rocks at the waterside," and then, in summation, "finally, not to leave any part of this great Lake that we have not explored, we are assured that in the interior, toward the south, mines of this metal are found in different places."[25]

This lengthy account, undoubtedly drawn from careful and critical listening to many sources, is amazingly complete and accurate, and the picture he presented was not significantly improved on for a century and a half. The reference to the large copper boulder on the Ontonagon River is clear, and the "verdigris" is undoubtedly the famed "roche verte," or green rock vein, found near the tip of the Keweenaw peninsula and which was to attract much attention in future years.

The *Relations* for 1670-71 also included the famed Jesuit map of Lake Superior (fig. 3). This map, an amazing cartographic achievement, was of unparalleled accuracy for its time.[26] Although his main purpose was his mission work, Dablon nevertheless noted that by glancing at his map "the reader will also be enabled—on his journey, so to speak—to note all the places on this Lake where copper is said to be found." These evidences, in turn, "all seem to force upon us the conviction that somewhere there are parent mines which have not yet been discovered."[27] Dablon's reasoning abilities were clearly not limited to the spiritual world.

With this firm information as a solid base, France began a century of sporadic but unsuccessful efforts to establish a profitable copper mining enterprise on Lake Superior. These attempts were to be compromised by great difficulties of transportation, repeated Indian hostilities, and, ultimately, a lack of enough demand for the metal to make the efforts profitable. For the French, from the time of their first contacts with the New World, the discovery of mineral wealth that would enrich the home country had been the long-range hope. The Jesuits were part of the eyes and ears of that effort.

Jean Talon, then the intendant of New France, had received a large mass of copper from Dablon in 1667, and he was anxious to establish active mining. He dispatched Jean Peré, an experienced voyageur, to investigate, but when he did not return Talon sent Louis Jolliet with additional supplies. Two priests, Dollier and Galinée, met Jolliet on his way to the northwest and wrote that he "had orders from the Governor to go up as far as Lake Superior to discover the situation of a copper mine, specimens from which are seen here that scarcely need refining, so good and pure is the copper." Yet, nothing came of the effort. The transportation difficulties seemed prohibitive, and Talon's recall to France stopped further efforts for a decade.[28]

From about 1680 to 1710, the district continued to garner comment without commitment. La Tourette, active in the fur trade in western Lake Superior, in 1687 brought "a large ingot of copper" to New France. About the same time Baron Lahontan wrote that "upon that lake we find Copper Mines, the Mettal of which is so fine and plentiful that there is not a seventh part lost from the Oar [sic]." Le Sueur, another trader, went to Paris in 1698 to get permission to try mining. Indian hostilities were a constant complicating factor, however, and the intendant was not impressed, saying that the only mines Le Sueur was interested in were mines of beaver skins.[29]

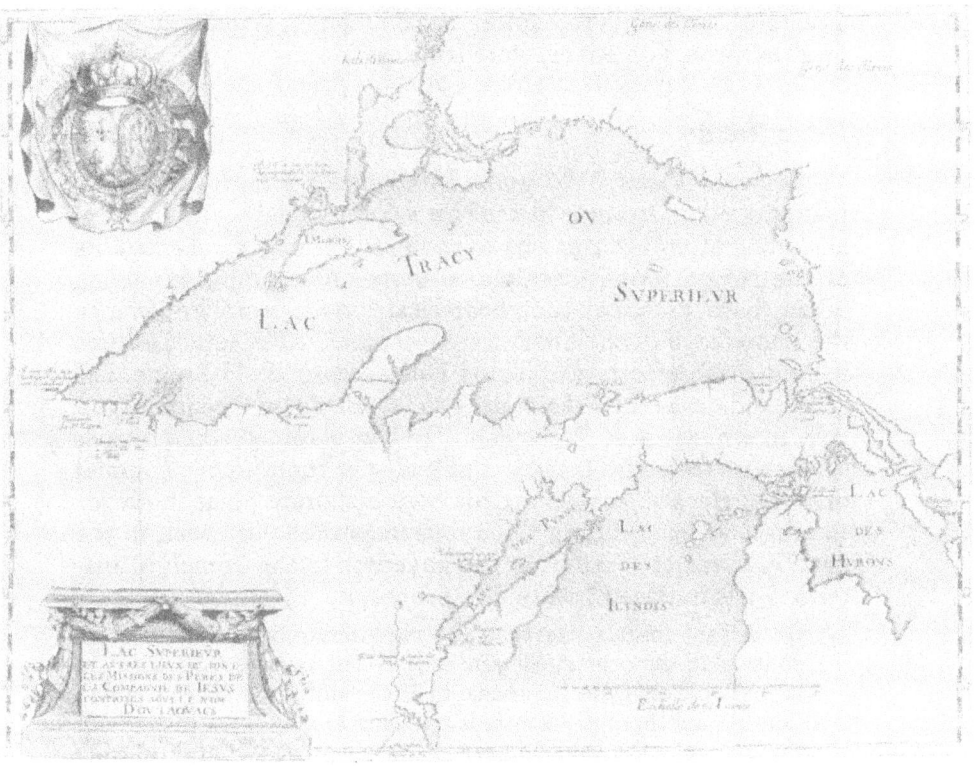

Fig. 3. *The Jesuit map of Lake Superior of 1670–71, published in Paris in the Jesuit* Relations *in 1672. The amazing accuracy of this map may be seen by comparing it with the modern map of figure 2 and with the Carver and Schoolcraft maps of figures 4 and 10 that represented the state of knowledge, respectively, a century and a century and a half later. Courtesy William L. Clements Library, The University of Michigan.*

By 1710 a new vision had taken hold in New France, and in a report the intendant wrote: "It is almost certain that there are copper mines on the borders of this lake and in the islands within its extent. There are found in the sands pieces of this metal, which the savages make into daggers for their own use. Verdigris rolls from the crevices and clefts of the rocks along the shores, and into the rivers which fall into the lake."[30] The ministry then declared the mines to be of high priority, but once again Indian wars intervened for nearly two decades. The governor's attempts to explore the deposits, particularly those on the Ontonagon River, were fruitless.

De la Ronde

Between about 1725 and 1740 a major French effort was inaugurated to attempt mining on a commercial scale. This episode has often been passed over lightly in historical accounts of the region, but it resulted in some important steps being taken, both practical and theoretical, which were to advance the knowledge of the district significantly.[31] The moving force behind these efforts was Louis Denis, Sieur de la Ronde, the commandant at the French post at La Pointe on Chequamegon Bay, in what is now Wisconsin. He had obtained some copper specimens from the Indians and had sent them to the colonial officials. He also sent along his own elaborate plans, both for mining and for building ships to transport the ore back to the East. A letter of 1732 from the governor to the French Minister in Paris outlined La Ronde's proposal:

> All the Information that we have hitherto been able to obtain consists in the continued assurance of The savages that the mine is in The Tonagaun [Ontonagon] River and in the Rivère Noire [Black], and there is a constant tradition among them that one of the islands in the Lake 25 Leagues from land, is full of this metal; that the block of this mineral lying on the shore of the Lake to The West [at the Iron River, in Wisconsin] comes from one of those islands, and was carried there by the ice.
>
> On this Information which is more than mere conjecture, Monsieur De la ronde proposes, in the memorial which we have The honor to annex hereto; to build two barks at his own expense: one on Lake Superior and the other on Lake huron [sic] or Michigan; in order that, by means of these barks, he may discover the islands in question, load them with copper from the mine and transport the same to Niagara, whence it can easily be taken to Quebec.
>
> We consider The Sieur La Ronde's project a bona fide one.[32]

Louis XV, ever on the alert for projects that would enrich France and enhance its standing in the eyes of the world, became directly involved in the endeavor. He carefully noted the communications sent to France about the colony, and particularly the possibility of mining the copper. After reading la Ronde's proposal, Louis explicitly placed his stamp of approval on the attempt to establish an active mining enterprise.[33]

The next year a piece of the copper was sent to Paris for assay, and evidently some doubt was expressed there that the copper was in fact a natural ore. The governor replied to the French minister's inquiry for further information:

> On Examining The Ingot of copper that we sent last year, one might be inclined to believe that it had been cast; And If it has been acknowledged by Monsieur Grassin [the assayist, presumably] as a work of nature, the pieces we send this year will convince him still more, Independently of the certainty we have that these specimens were cut with hatchets from larger pieces of the same quality. We consider that the whole difficulty in the Working of these mines, supposing that the ore is as pure as it seems to us to be, Consists at present in finding means to divide into portable pieces, the masses of Copper on the banks of the river Tonagan [Ontonagon], and that which will no doubt be also found at other Places on Lake Superior. . . . This is doubtless the reason which has Induced the Sieur de la Ronde to write us to beg you to procure a skilled workman for him to show them how to Work these mines, which, he says, consist only of a Mass of ore. . . . It seems to us that he really intends to prosecute this Undertaking.[34]

A carpenter named Corbin, who worked on the vessel built at the Sault for use on the lake, traveled to the mine at Ontonagon. In his report, which was also sent to the French minister, he claimed to have seen fifty unbroken layers that contained copper, all bound together by sandy soil without stones.[35] He was a carpenter and not a miner, but the Ontonagon Boulder did come through in his report as a mass "on the bank of the said River Tonnagane at the foot of the mine which seems to have rolled down from the Cliff, and which is from 6 to 7 feet in Diameter.'

Long frustrated by the success of other nations in exploiting the New World, the French officials were beginning to warm to the subject: "If The affair of the copper mines should have as advantageous results as We have Reason to Hope, this Country will become more and more worthy of his Majesty's attention and of the jealousy of our neighbors."[36] De la Ronde did his part by showing up in August 1736 with five hundred pounds of copper that had been taken from the masses on the Ontonagon and Iron rivers. Samples, one weighing 110 pounds, were carefully classified, labeled, and sent to Paris. He also tried to reach "the island where the virgin Copper is said to be," but storms prevented his landing.

De la Ronde continued to press for someone with mining experience who could assist him. He had no background on which to rely in his search for copper: "Had I had any knowledge of Minerals, I should certainly have had no need of Miners," he wrote. In response, two German miners, John Adam Forster and his son, were sent to Lake Superior to aid and advise him. La Ronde missed his connections with them at the Sault, and they went on to make their first tour of the southern shore without him, aided by an Indian guide. He finally met them when they returned to the Sault rapids again, and their discussions soon revealed that they, as experienced miners, and he, as an inexperienced entrepreneur, had different opinions on the significance of what they had seen.

De la Ronde, like so many before him, had simply assumed that the importance of the copper was self-evident. The masses found scattered over the countryside had almost always been regarded as "mines" in and of themselves. It was thought that such masses were just lying there, clearly visible, waiting to be carried away by someone with enough ambition and enthusiasm, but the Forsters did not agree. He asked them for their opinion of one of the large copper masses, and they replied that it had probably been "carried by the ice to that spot on the shore."[37] As they gave him a quick lesson in the basic principles of geology and mining, la Ronde began to realize that mining was more complicated than he had thought:

> They told me that they had found nothing but nodules of Copper (this is what we call Masses), but no main lode. I told them that they must re-embark with me. Thereupon they said that *Copper was not found in earth but in Rock* [my emphasis]. I replied that they had come from too great a distance to allow of their returning so soon; that I would find Rocky bluffs for them in the neighborhood where we could certainly discover the main lode; which I did.[38]

They pointed out to him, for example, that the masses found in the clay and sand at the Ontonagon "were only fragments that had come from the Mountains." Gradually it dawned on La Ronde that all that he had seen thus far were only surficial indications of what must lie beneath the ground. The real riches, if present, would of necessity have to be found in the solid bedrock of the earth. The Forsters were evidently anxious to get back home, but La Ronde insisted that he would pay them well for the extra time, so they agreed to set out again.

Fortunately, he was not easily discouraged, and he learned fast. Soon he was echoing them, speaking of finding veins of copper in the rock itself on both the Ontonagon and St. Anne rivers. La Ronde said that the miners assured him that the second location was "as good as any mines in their country," and he determined to make that location the seat of the new colony he was planning.[39] His vision of the future then expanded dramatically. He saw "miners, founders, Carpenters, and a blacksmith." Evidently impressed with the background of his advisors, he wrote that the new workers "must come from the Mines of Germany so that they may be thoroughly conversant with what they have to do," and there will be "Barracks, good Magazines, and blast furnaces of the German System." As soon as the cattle arrive, "the Colony will be flourishing; for there are no better lands nor meadows throughout Canada . . . while the climate is very mild." His enthusiasm was apparently contagious, and it spread throughout the country. Apparently even some of the advance money from the royal treasury was reimbursed by the sale of the copper sent east. The future, seemingly, was bright.

The report of the Forsters, however, was not quite as optimistic as La Ronde, in his enthusiasm, made it out to be. At the Ontonagon,

> they found only one piece of rock [the Ontonagon Boulder] from said mine which could truly contain a thousand pounds of copper; besides, there did not seem to be any absolute indications of a mine at that place; but in returning toward Lake Superior, at a distance of a league and a half from there, they found a vein or lode from which this piece could have been taken, as the vein contained a little pure copper in the matrix.[40]

Of the two prospects farther west, at the first they found "a vein or lode where copper can be recognized in the matrix and which is very hopeful," and at the second there was "a good mine of coppery slate in one layer only." Not as promising, perhaps, and they were wary of the great distance that the ore would have to be transported. La Ronde, perhaps not wanting to hear too many discouraging words, paid the Forsters for their efforts and sent them on their way back to Europe.

It was, unfortunately, just too much, too soon. An important, if sobering, step had been taken with the realization that long-term mining success could not be based just on copper masses lying around on the surface of the land. Instead, copper-bearing minerals would have to be found "not in earth but in rock." The prospects, at the time, for adequate supplies of any ore were simply not great enough, nor was the demand for the metal enough to ensure profitability, even if an abundance of ore had been in sight.

As the Indian wars began raging again, la Ronde's plans began to disintegrate. Pleading his loyalty to the crown to the last, he was left to appeal to the generosity of the French minister for the well-being of his wife and the royal favor for his children. Dying soon thereafter, he left his widow to survive on the largess of the government, based on the memory of the expenses he had incurred in attempting to open the mines. It was a significant venture, and one from which much had been learned, but it was, ultimately, simply born before its time.[41]

The French years in the district were now numbered, but the usual string of travelers continued their comments on the copper. Peter Kalm was a Swedish naturalist and a keen observer of nature who traveled in America from 1748 to 1751. In 1749 he was given a piece of copper from Superior while visiting at Montreal, and his brief comments were virtually a summary of what was known at the time:

> They find it there almost pure, so that it does not need melting over again, but is immediately fit for working. . . . One of the Jesuits at Montreal who had been at the place where this metal is native told me. . . that there are pieces of pure copper too heavy for a single man to lift up. The Indians there say they formerly found a piece about seven feet long and nearly four feet thick, all pure copper. As it is always found in the ground near the mouths of rivers, it is probable that the ice or water carried it down from a mountain; but, notwithstanding the careful search that has been made, no place has been found where the metal lies in any great quantity but only in loose pieces.[42]

The British Period

As the years passed and the Sioux-Chippewa wars kept things in turmoil in the country around the lakes, the always tenuous hold of France on its possessions in the New World continued to slip. Corruption was everywhere evident in the French administration, and the influence of the English became more and more pervasive.[43] Then, "the greatest turning-point as yet discernible in modern history," as one historian has characterized it, occurred with the capture of Quebec by the English in 1759.[44] "So we are gone" said the French minister, "it will be England's turn next." An empire, along with its attendant problems, changed hands as the Treaty of Paris of 1763 ceded to Britain all French territory east of the Mississippi River. This included the copper district.

The British held the copper region for a rather brief period. In taking control of the western posts of their newly acquired lands, they heard firsthand of the copper efforts of the French. In the pursuit of furs, the English trader Alexander Henry became one of the first to enter the lakes region following the British conquest, and he also carried out the initial British attempt to mine the copper. His well-known effort has been called "the first mining operation conducted by Europeans" and "the first recorded attempt by white men to exploit the fabulous riches on Lake Superior's shores,"[45] which it clearly was not, in the light of the earlier work by the French.

At Michilimackinac, the English fort at the straits between Lakes Huron and Michigan, Henry achieved some degree of fame by barely escaping with his life from the famous Indian massacre. On passing the mouth of the Ontonagon River during his travels in 1765, he saw copper masses in the hands of the Indians, and on his return the next year he paused to go upriver with Indian guides to view the great copper boulder: "The object, which I went most expressly to see, and to which I had the satisfaction of being led, was a mass of copper, of the weight, according to my estimate, of no less than five ton. Such was its pure and malleable state, that with an axe I was able to cut off a portion, weighing a hundred pounds."[46] In the meantime, a group of speculators and traders was giving increasing attention to the copper stories. Robert Rogers, Henry Bostwick, and Alexander Baxter, Jr., had enlisted the support of prominent people in England as a result of the rumors filtering back across the ocean. To set the stage for a mining effort, in 1760 Rogers had managed to negotiate a treaty with the Indians that granted him twenty thousand acres on the south shore of Lake Superior, part of which he sold while in London. Sir William Johnson, a partner in the endeavor, reported that any attempt at mining would encounter grave difficulties, particularly regarding transportation of the ore, and General Gage, commander of the British forces in America, doubted that the mines could be made profitable unless gold was found with the copper.[47]

Nevertheless, a company was formed and the building of appropriate vessels was begun on the same point near the Sault that De la Ronde had used earlier. In 1770 Baxter returned from England bearing the organizational papers for the company, and Henry was made a "joint-agent and partner" in the venture. On one of his exploratory trips along the south shore of Superior just west of the Sault, Henry took along a Mr. Norburg, "a Russian gentleman acquainted with metals," who was stationed with the British garrison at Michilimackinac. Norburg happened upon a loose stone "of eight pounds weight, of a blue color, and semitransparent." Carried back to England, it was, according to Henry, placed in the British Museum after analysis revealed it to be 60 percent silver. Now the rumors had even more fuel.

First, in the spring of 1771 Henry and his miners tried to find gold on the "Island of Yellow Sands," the present small, isolated Caribou Island near the eastern end of Lake Superior.[48] Failing in that endeavor, they sailed again, this time heading westward to the Ontonagon River and the site of the great boulder where, besides the frequent detached masses of copper, they claimed to also see "much of the metal bedded in stone."[49] Believing it was time to begin mining, and apparently with absolutely no forethought or geological insight, they decided to simply drive an adit, or tunnel, into the side of the clay bluff near the boulder, hoping to find more of the metal. Having instructed and equipped the miners for a winter's work, Henry departed back to the Sault.

The next year, in the spring of 1772, Henry sent a boatload of provisions to the miners, but it soon returned "bringing with it, to our surprise, the whole establishment of miners." In their search for the metal they had penetrated forty feet into the clay bank, which then collapsed with the first thaw, and the enterprise was abandoned. Even if they had found copper, Henry then rationalized, it would have cost more to transport it to Montreal than it was worth. Anyway (he now said), the company had always hoped for silver, rather than copper. That summer the group dabbled at mining on the north shore of the lake, sending some ore to England, but by 1774 "Mr. Baxter disposed of the sloop, and other effects of the Company, and paid its debts." Like that of the French that had preceded it by thirty-five years, the great English copper-mining venture, a largely hit-and-miss operation, had failed.

Even so, out of this period there came a publication that was to play a large part in keeping the promise of the copper mines before the public eye, both in this country and abroad. Jonathan Carver was an associate of Rogers who had spent much of 1766 and 1767 exploring the region of the upper Mississippi River. In 1778 he published his account of those years as *Travels through the Interior Parts of North America*. Although it revealed him to have been, in many respects, "either a gullible innocent or a liar, probably both," the account achieved immense general and even international popularity. It went through many editions, was translated into several languages, and became "a part of every respectable library around the world."[50] Of particular interest is the map that Carver included with the work (fig. 4).

On his journey Carver had traveled along the north shore of Lake Superior, and therefore he never saw the copper region. On his map he included fictitious islands and mistakenly identified a greatly deformed Keweenaw peninsula as "Point Chegomogan," confusing it with the point at the base of the Apostle Islands that was another hundred miles farther west toward the head of the lake. Nevertheless, perhaps at the suggestion of Rogers, across a large river clearly meant to be the Ontonagon, Carver wrote, "About here is Plenty of Virgin Copper." This became the first printed map to localize the native copper to the western portion of the south shore of Lake Superior. These same words or closely related phrases about the copper were then included on new maps of America that were increasingly appearing about the time of the Revolution. They helped ensure that the lure of the copper would remain alive as the region was about to undergo a final and far-reaching change of jurisdiction.

By this time nearly three centuries had come and gone during which two major European powers had made attempts to investigate and exploit the copper district on Lake Superior. All those efforts had been sporadic and inconclusive. The fate of the region, however, was now to pass into other hands. During the war for American independence the land along the southern shore of the lake had been only a far outpost of the Canadian territory. It saw little of the revolutionary war action that dominated the colonies to the east, and it remained under firm British control at the end of the war.

For some time the direction that the future of the region would take was not at all clear. At the peace negotiations in

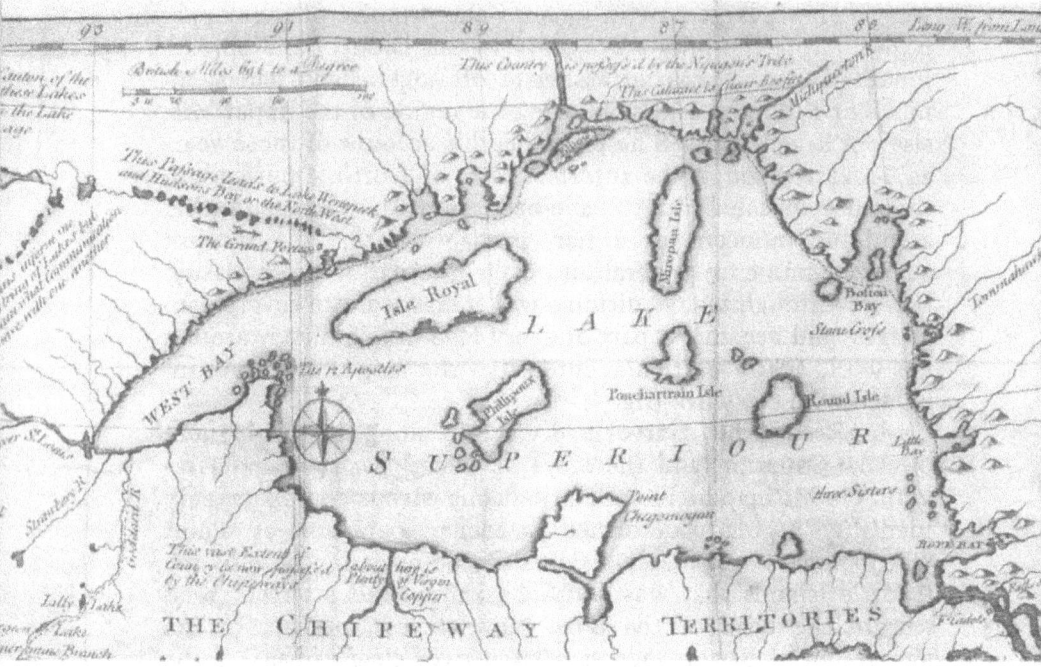

Fig. 4. Detail of the Carver map of 1778 based on his travels in 1766–67. Its accuracy can be compared with that of the Jesuit map of a century earlier (fig. 3). Note the statement near what is intended to be the Ontonagon River, "about here is Plenty of Virgin Copper." Courtesy William L. Clements Library, The University of Michigan.

Paris one proposal would have made the forty-fifth parallel the northern boundary of the United States, in which case all of what is now the Upper Peninsula of Michigan, and the northern part of the Lower Peninsula as well, would have become permanently a part of Canada. The Treaty of Paris of 1783, however, finally set the boundary at the middle of Lakes Ontario, Erie, Huron, and Superior, and thus the copper district became a part of the vast western lands of the United States.

A persistent, long-lived claim has been made by some that the existence of copper on Isle Royale was a factor in the negotiations between the parties at the peace conference. According to this story, Benjamin Franklin's awareness of that copper led to his insistence that the international boundary be bent northward so as to include that island within the territory of the United States. This attractive story is sometimes told with elaboration, even quoting conversations between Franklin and various Europeans, but unfortunately it appears to have no documentary support.[51]

In reality, the British presence on the lakes continued to remain strong for some years after the Revolution, and it was not until 1796 that Detroit and the fort at Mackinac were finally turned over to the United States. The Keweenaw district now became officially American, and its future, whatever it was to be, had passed out of the hands of the Europeans. With the opening of the new century, the district was soon to find itself embraced by the vigorous drive to the west of the new and expanding country.

The beginning of the nineteenth century introduced an ever increasing interest in the copper region of Lake Superior that culminated, shortly after midcentury, in the establishment of the district as one of the great mining areas of the world. To see these events in their proper perspective, and to understand why some parts of the story have been misunderstood and misinterpreted for so many years, it is necessary to digress slightly to examine two important preliminary questions. First, *how* and *why* did the copper found in the Keweenaw differ from the familiar copper ores that were being mined in other parts of the world? Second, exactly what were the persons who began exploring the Keweenaw during this period *expecting* to find? Only with this information as background can the questions and uncertainties of the following years be understood, as the value and importance of the native copper of the Keweenaw was gradually revealed.

2

Amateurs, Professionals, and Native Copper

The Creation of the Keweenaw

The Keweenaw peninsula is a part of the most ancient portion of the North American continent, and the events that formed its rocks and the copper found within them began far back in geological time. About one billion years ago the crust of the earth in that area was trying to split. Huge moving currents of dark, molten rock rose from the interior of the earth, and as they pressed upward toward the surface they forced their way through lines of weakness in the earth's crust. These fractures were apparently located beneath what was to become, only very much later, the western end of Lake Superior. The molten rock poured out layer upon layer, most of which upon solidifying formed a dark lava called basalt, which is also often referred to as "traprock" or simply "trap." At least two hundred lava flows of this type, perhaps many more, were successively poured out onto the surface, and some of them were of tremendous size. One of these, the Greenstone flow, was in places over a thousand feet thick, extended over fifty miles, and contained more than two hundred cubic miles of rock; it may be the largest single lava flow on earth. This thick sequence of flows, now called the Portage Lake Volcanics, includes the rocks within which most of the native copper was later found.[1]

Other geological processes were also operating in those early times. Occasionally there were pauses in the eruptive activity that produced the lava flows, and farther to the southeast there were higher lands composed of different kinds of rock. Streams and rivers flowed off these lands, carrying boulders, pebbles, and sand of a distinctive reddish color that were then deposited as layers of gravel on top of the lavas. In some places these layers were nothing more than a thin seam, but in others they were as much as several tens of feet thick. These pebbly layers were then buried by the next lava flow, and the process was repeated. In time, these gravel layers solidified into rocks that are now called "conglomerates." The Portage Lake Volcanics, therefore, is not made up entirely of lava. It also includes conglomerate layers now found sandwiched between the much more numerous flows.

Later, strong forces acted to squeeze the region together, as if in the jaws of a giant vise, and the region took on the form of a large basin, or "syncline" (fig. 5). The outer, upturned edges of the lava flows and conglomerates now form the northwestern half of the present Keweenaw peninsula, where the layers dip out under Lake Superior. A major fracture, the Keweenaw fault, now separates these rocks from the more flat-lying and probably younger Jacobsville sandstone rock that makes up the southeastern portion of the peninsula. Almost as if seen in a mirror, the Portage Lake Volcanics are also found on Isle Royale, where the same rock layers dip the opposite way, southeastward out under the lake.

The Copper Lodes

The native copper deposits of the Keweenaw were found in three distinct forms called *lodes*.[2] One was the amygdaloid lode, in which native copper was found as one of the minerals that filled what formerly were empty gas-bubble holes and small fractures that had formed in the upper portion of some of the lava flows as they had cooled. The second was the conglomerate lode, in which the copper formed part of the material that filled the spaces between various pebbles and grains, or replaced some of the minerals, of the rock layers that had formed from the gravels laid down by the streams and rivers. In both lodes, the copper was found mostly in small particles scattered throughout the rock, occupying what were originally small empty spaces.

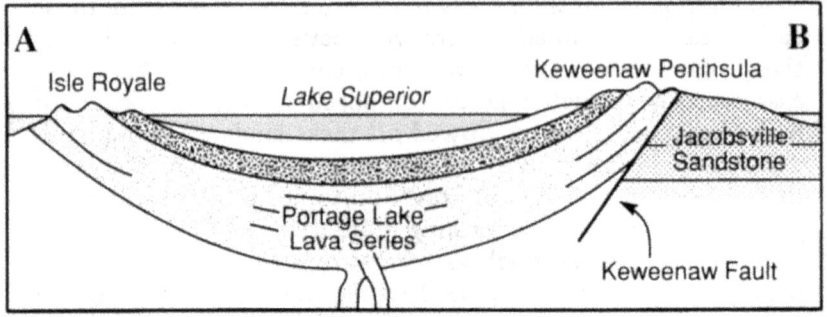

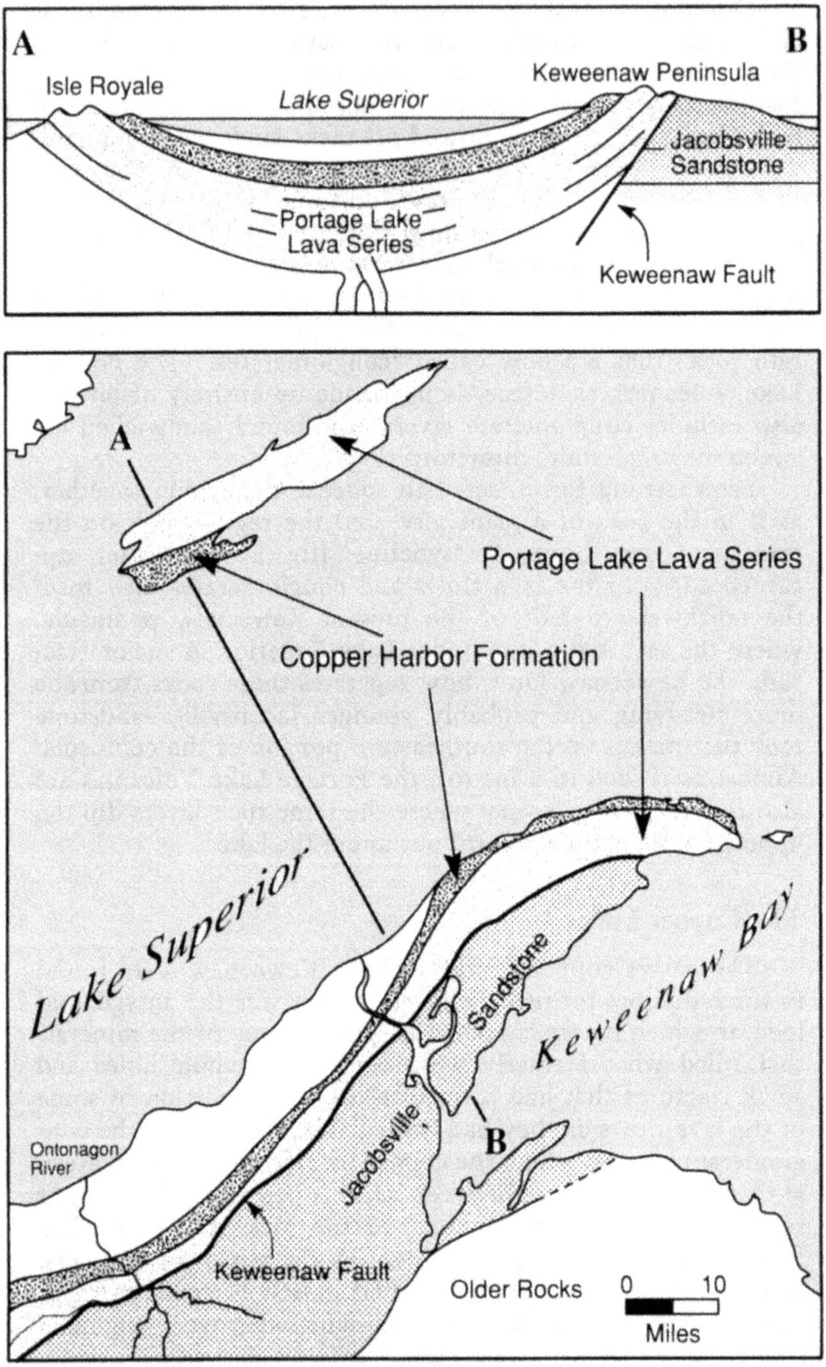

Fig. 5. The geology of the Keweenaw peninsula and Isle Royale, showing the most important rock units. At the top is an idealized cross-section along line AB.

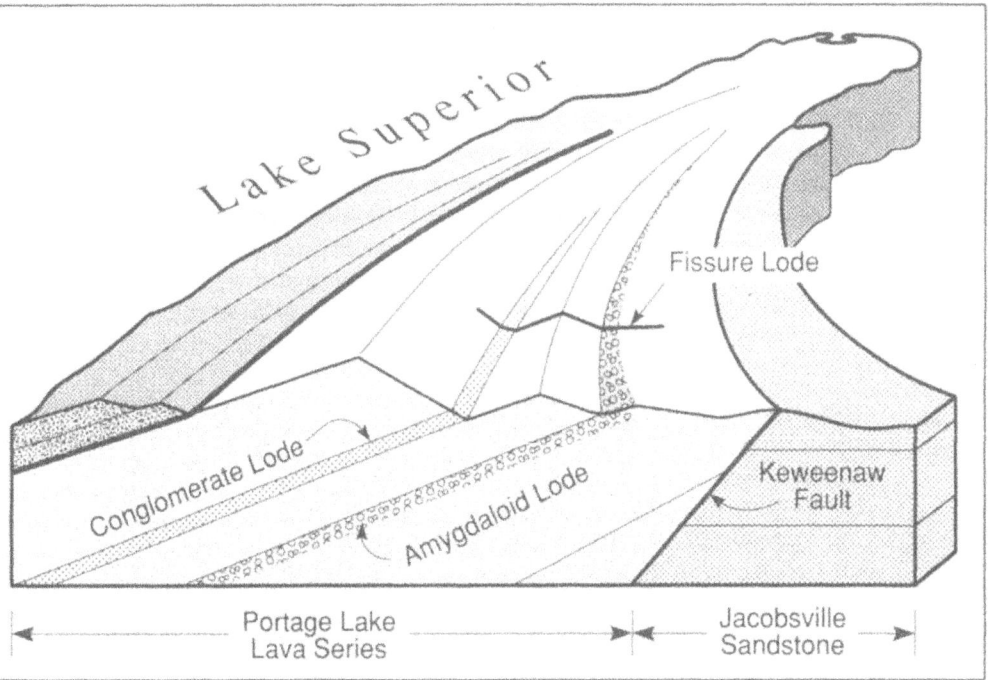

Fig. 6. Simplified sketch of a cross-section of the Keweenaw peninsula looking north, showing the rock units and the three copper lodes: the fissures, amygdaloids, and conglomerates.

Because of the downwarped structure of the district and the effects of vast ages of erosion, these two lode types appeared at the surface of the earth in narrow but elongated stripes that followed the rock structure of the district from the southwest to the northeast (fig. 6). It is important to note that these two lodes conformed exactly to the rock structure of the region, running parallel to the rock layers at the surface and dipping at the same angle with those layers as they went into the earth, downward and outward toward Lake Superior. Well over 95 percent of all the copper produced in the district eventually came from these amygdaloid and conglomerate lodes, but of the hundreds of lava flows and perhaps dozens of conglomerates layers interbedded between them, only a very few were found rich enough in copper to be mined profitably.

A third type of lode was also found, and it was to be of great historical importance: the fissure lode. When the rocks of the district were originally forming, numerous fractures or fissures formed that cut across the other rock layers, probably as one result of the pressures that were deforming the earth's crust. Many of these fractures showed only a few tantalizing traces of copper-bearing minerals. Others of them, however, contained masses of pure copper that were much larger than those normally found in the amygdaloid or conglomerate lodes, and some of these masses were truly immense in size. They were Keweenaw copper in its most spectacular form.

In contrast to the amygdaloid and conglomerate lodes, however, most of the fissure lodes did *not* follow the usual trend of the rocks. Instead, they cut across the rock structure, frequently at nearly right angles, and most of them also dipped almost vertically into the earth. It is important to realize that even though the fissure lodes were very few in number and contained only a very small proportion of the total copper in the region, throughout the years that led up to and included the opening of the district to mining they were the *only* lodes known or even suspected to exist. The fissure lodes were to have an importance that far exceeded the simple dollar value of the copper they contained; the amygdaloid and conglomerate lodes, although they contained vastly more copper, were to be later arrivals on the mining scene. Despite their individual differences, however, these lodes *all* contained native copper rather than other copper-bearing minerals, and it was this that contrasted so greatly with the other copper-mining regions of the world. These contrasts will be examined here briefly.

Copper and Its Ores

For any copper-bearing mineral to be called an *ore* of copper, it must be found in sufficiently large and accessible amounts so that it can be profitably mined and processed.[3] Copper is a metal that has a strong chemical affinity for other elements, and it is almost always chemically combined with either sulfur or oxygen when it is found in minerals in the earth. Copper ore minerals are therefore commonly classified as either sulfides, oxides, or perhaps carbonates, which contain both oxygen and carbon. At the opening of the Keweenaw district, the most commonly mined copper ore mineral of the world was chalcopyrite, or the "yellow copper ore." This one mineral, consisting of a combination of copper, iron, and sulfur, furnished nearly all the ore of the famous Cornwall district of Great Britain. It also provided about two-thirds of all the copper produced in the world at that time, and today it is still the single most important copper ore mineral.

Even when examined closely, however, such minerals do not in any way resemble metallic copper. After being mined, such ores must then be smelted so that the metal can be liberated from the other elements that are part of the mineral. Smelting is always an involved procedure, and in the early 1800s the processing of sulfide copper ores required as many as ten distinct and separate steps before the metal was reasonably pure. Typically, of the material actually removed from a mine, less than 10 percent would emerge at the end of the smelting process as metallic copper, and in many mines the copper content of the ore might be significantly under 1 percent by weight.[4]

In the 1840s, just before the opening of the Keweenaw district, the copper mined in this country totaled only about one hundred tons per year, mostly from small producers of ores of copper compounds in Eastern states. This was far below the needs of even that day, and the United States was forced to import several thousand tons of refined copper each year.[5] By 1800 Britain had become the world's largest copper producer; most production came from the mines of Cornwall. The city of Swansea in Wales was the world center of copper refining, and the Metal Exchange there was the hub of the world copper trade. Ores of copper from around the globe poured into Swansea for processing, where the required smelting skills were tightly controlled by a few companies.

Exactly how were these ores of copper different from the native copper of Lake Superior? In some historical accounts, authors have made extreme statements implying that native copper was almost totally unknown in the world before it was found in the Keweenaw. One recent writer, for example, has stated that when the pure copper was found there, it "sorely perplexed the geologists of the time, who had never before encountered the metal in its pure form in nature."[6] Statements such as this, however, are clearly not accurate. In reality, native copper is not a particularly rare mineral. Even before the Keweenaw had been explored, native copper had frequently been found in many other places in the world, often where other ores of copper were being mined, and occasionally even in very large masses.

Even granting this, however, what *is* true is that in these other regions the total amount of native copper was always small if compared to the other copper-bearing minerals with which it was always associated. Such native copper had formed when weathering processes at the surface of the earth acted on the other minerals present; in effect, nature had done its own small-scale smelting by freeing the copper naturally from the other elements with which it had been combined. Such native copper, however, was found only near the earth's surface, and it would disappear as mines were pushed deeper in order to reach the more valuable minerals, usually copper sulfides, that would be found in large quantities at greater depths. Because under such circumstances native copper was clearly only a secondary, relatively unimportant mineral, by the early 1800s it was generally agreed, by those most familiar with minerals and mining, that native copper was *never* found in quantities great enough to be of any value for mining. At best, it was thought to be only a possible clue that larger quantities of sulfides or other copper-bearing minerals might also be found at greater depths within the earth. This view was to have far-reaching consequences in the struggle to come to grips with the true nature of the native copper of the Keweenaw.[7]

Why is it, then, that the copper of the Keweenaw was all found in the pure, native condition, though in virtually all other mining regions of the world the ores are almost all composed of copper chemically combined with other elements? All of the details of the processes that must have been involved are still not completely understood, and the question of exactly how ores form is still being intensively studied today. However, a simplified scenario for how the native copper of the Keweenaw formed might be roughly as follows. After the lava flows and conglomerate layers of the region had been deposited, they were buried very deeply by younger, overlying layers of other kinds of rock. At those great depths the pressures and temperatures became high enough that water within the lavas dissolved out minute amounts of copper. The pressure then pushed this water outward and upward through the small openings in the amygdaloid and conglomerate layers. As it rose and cooled, the water then apparently redeposited the copper in the empty spaces and fractures in the rocks closer to the surface, where the pressures and temperatures were lower. Because there was little sulfur in the rocks, the copper was deposited in the pure, native form, rather than in the form of the sulfide minerals so common in other regions. It appears, then, that the unique native copper deposits of the Keweenaw did not come into being as a result of any particularly unusual or exotic chemical reactions or conditions. Instead, it would appear that a set of fairly straightforward geological processes acted in combination to produce, in this small corner of the globe, a result so distinctive that its like has not been found on such a scale anywhere else on this earth.[8]

After these geological events had formed the rocks and their associated copper riches, the area became a part of the stable nucleus of what is now the continent of North America. For some six hundred million years, during which time life developed into its varied and diverse forms, mountain ranges rose and were eroded away, and continents were joined and divided, the district lay relatively quiet. It was on some occasions submerged beneath warm tropical seas and at other times part of a vast, eroding land, but it was largely removed from the major geological events elsewhere that were molding and changing the face of the earth.

Then, as the Ice Age dawned perhaps only a million years ago, the climate grew colder. In the North, the ice began to accumulate and to push its way southward. As it moved over the land where the copper was, it plowed away much of the softer rock that flanked the Keweenaw district on either side. The upturned edges of the more resistant lava flows and conglomerates were left standing as ridges that, smoothed and rounded, pointed to the northeast like a long, crooked finger above the surrounding softer rocks. Finally, with the last retreat and melting away of the final ice sheet only some ten thousand years ago, the lowlands on either side became filled by the greatest single body of fresh water on earth, Lake Superior, but the ridges of the district were left standing as a long, narrow peninsula pointing toward the center of the huge new lake.

As the ice disappeared it covered much of the bedrock that lay beneath it with the residue it was carrying, concealing most of the copper lodes under a blanket of soil, sand, and mud. However, the moving ice had torn some of the copper from the veins and crevices in the rock that had been its resting place for a billion years. These "float" masses, left scattered haphazardly across the countryside, were the tantalizing clues that were so important in drawing early attention to the region. One particularly large mass of copper came to rest near what later became the bank of a large river that, miles downstream, flowed into the great lake. It was the famed boulder on the Ontonagon River.

As the years passed, prehistoric Indians hacked and gouged at these masses, removing any projections of copper and, leaving them with a beaten and flattened appearance (fig. 7) and also, if rumor is to be trusted, conducted various religious rituals on the largest of them. The Indians also discovered the copper metal outcropping in the bedrock, and in many places they carried out mining on a small scale, leaving the peninsula and the great island far out in the lake dotted with shallow rocky pits and their hammerstone tools. The copper was fashioned into decorative items and useful devices that became widely distributed as they were traded over long distances by the simple commerce of the times, and were the artifacts that so early caught the attention of the Europeans.

Nevertheless, even after all the efforts of the Indians and the nearly three centuries of attention by two major European powers, almost all the copper, its surface barely scratched, still lay undisturbed and unsuspected, deep within the rocks beneath the surface. Now, however, as the nineteenth century dawned, a period of greatly increased interest in the copper region was about to begin. To this point, what exactly had been learned about its nature and what was as yet not clearly understood?

Considering the sporadic and unsystematic efforts and the difficulty of the circumstances, some significant steps had been taken. An unusual mineralogical condition on an unprecedented scale had been recognized on the south shore of Lake Superior. Furthermore, the masses that were found so frequently scattered over the surface around the western end of the lake had been recognized for what they were: only surficial fragments not likely present in any locality in amounts great enough to mine. In addition, the attempts at mining that had been made had resulted in an increased appreciation of the need to find veins or beds of the metal within the bedrock, if long-term success was to be achieved. Finally, such deposits in the rock were recognized on Isle Royale and were suspected to exist toward the interior part of the district along the southern shore of the lake as well.

Yet, despite all this, the true value of the native copper was a long way from being clearly understood. Much of what

Fig. 7. Large mass of native copper from Isle Royale, weighing nearly three tons and showing the effects of having been worked by ancient Indian miners. The mass was photographed on a glass plate on the docks in Detroit in 1874 and was later melted down after being exhibited at the Philadelphia Exposition of 1876. Courtesy Burton Historical Collection, Detroit Public Library.

was known was still based on rumor, hearsay, or second-hand telling, and much of what the French had learned lay buried in governmental files. More importantly, a real grasp of its value required much more than the simple recognition that great amounts of pure, native copper were present. This straightforward fact had seemed apparent, and perhaps even obvious, to a great many people for many years. The real heart of the issue was simply this: what did the presence of this native copper *mean*? One reason that the coming events were more complicated than might, at first, be anticipated is that determining the true nature of the copper was to involve much more than just making accurate or careful observations. The barriers to a clear understanding were not always in nature alone, but they were instead, at least in part, within the minds of men.

Three persons dominated this part of the history of the district: Henry Rowe Schoolcraft, Douglass Houghton, and Charles T. Jackson. All are well known for their involvement in developing the copper region. Less well known is the exact nature of their individual contributions and their interrelationships. Some published historical accounts of their work are seriously deficient, because their attempts to understand the value of the native copper were not based on just the empirical evidence. Conceptual factors, whose effects have not been fully appreciated before, were also involved. Therefore, to grasp the full significance of their efforts, it is necessary to take a closer look at how the science of geology was practiced in America during the first half of the nineteenth century, when the Keweenaw was coming to have an increasingly important part in America's westward expansion.

Science—Amateurs and Professionals

Before 1800, and even into the first decades of the new century, science in America had been an unrestricted, democratic field. It was the time of the amateur, the common man, the public figure. Scientific studies, as a separate field of endeavor, were not then sharply distinguished from a broad interest in the natural world. Such an interest might be characteristic of any curious and observant person's outlook on life. The science appropriate to the expanding frontier environment of that time consisted, naturally enough, largely of discovering, naming, and classifying. The emphasis was definitely practical, not theoretical, and occasionally there was even an antitheoretical bias. A simple empiricism that emphasized fact gathering was the general order of the day. "Useless speculation" was often avoided in favor of a more direct, "objective," and "unbiased" contact with nature.

There is little question that the picture of the copper region that had emerged by 1800 had done so largely in the light of this general philosophy. Most of those who had made contributions to that picture were explorers, traders, priests, travelers, or government officials, people whose real interests had lain mostly elsewhere. Their curiosity had not, for the most part, been motivated by any intellectual or scientific concerns. They had no theoretical axes to grind, they did not belong to any inner scientific cadre or circle, and their facts were open and equally available to all. If the copper they found in the Keweenaw was naturally pure, then pure it was, and that was that. Based on investigations of this type, by the early 1800s the nature of the native copper was well on its way to being understood for what it was: fundamentally different from the ores found in the other copper-mining regions of the world.

During the early part of the nineteenth century, however, science began to undergo a dramatic change of both image and practice, and this was to have some far-reaching implications. The day of the amateur, the public figure, and the casual observer was passing. The professional scientist was beginning to appear, the person for whom science was a single-minded dedication as opposed to a pleasant diversion from the other concerns of life. Particularly crucial in these developments was the interval from 1820 to 1860, a time that has been called "the emergent period" of this process of professionalization, and the process has been called "the most significant development in nineteenth-century American science."[9] Those who had an active, long-term commitment to a scientific understanding of the world were gradually coming to see themselves as a group with common and distinctive goals, and science was beginning to be viewed from an increasingly less democratic perspective, both by its practitioners and by the general public.

Several important effects were to emerge as a result of this process. One was institutionalization; those who viewed themselves as scientists sought to establish organizations to formalize their professional relationships. Geologists were in the vanguard of this movement, and their efforts at organization were to prove particularly successful and long-lasting, for they culminated in the creation of the American Association for the Advancement of Science. As one essential aspect of professionalization, such organizations confirmed the status of both those on the inside and those on the outside. Scientific journals were beginning to reinforce the distinction between "real science" and its popular imitations, and they began to enforce more rigid standards on the scientific material that appeared in print. Without doubt the most important of these periodicals was Benjamin Silliman's *American Journal of Science*, occasionally referred to simply as Silliman's Journal. It has been called "the greatest single influence in the development of an American scientific community."[10]

Another effect of professionalization was an increasing emphasis on at least some educational background in the field as a prerequisite for membership in a scientific group. In the simple, fact-gathering stage of science, a sharp mind and keen observational powers might have served. Now that the question of who was a professional and who was not was becoming of greater concern, some evidence of a formal understanding of the basics of the field was considered desirable. Once again geology was in the forefront, and early institutions such as the Rensselaer School in Troy, New York, came to have a particularly recognized status. Already by 1840 the term "popularizer," with its "modern invidious connotation," began to be applied more freely to those who did not measure up. Attempts to make the *American Journal of Science* into a more popular public medium, for example, were pointedly resisted. A few scientists were even beginning to reject calls that their work should be capable of being understood directly by the public. One opined that "we cannot write anything for the scientific few which will be agreeable to the ignorant many," and even the notion of "popular science" was specifically attacked.[11]

It was common at this time to encounter the claim that if science would become a more professional enterprise instead of remaining the domain of amateurs, the American public would directly benefit from the technological applications of new scientific discoveries. If it was true, as James Dana, one of the most prominent of early American geologists, put it to the Yale alumni in 1856, that "science is an unfailing source of good," the taxpayer would be well advised to encourage and support (with their tax moneys) the new scientific professionals.[12]

One additional result of professionalization was to prove to be important in the attempt to understand the nature of the copper. Increasingly, theoretical considerations were being placed above the straightforward descriptions and classifications that had been the norm earlier. The "fact gathering" of earlier years had resulted in accumulations of vast amounts of undigested data, often with no apparent interrelationships. A science based on simple observation alone was exhausting its potential and, according to this new image, a scientist belonged to a united community, with an accepted institutional structure, a common educational background, and a commitment to a common theoretical framework. With the beginnings of such professional groups, judgments about the "bare facts" were increasingly being made by appealing to a backdrop of common traditions, the acceptance of which became an essential aspect of membership in the group.

One approach used by scientists in trying to organize the great amounts of raw data that had been accumulating was the "doctrine of analogy." This allowed similarities to be perceived and relationships to be established among data that otherwise might have been entirely overlooked. It is, to be sure, a most reasonable principle, and Amos Eaton, under whom Douglass Houghton was later to receive much of his geological training, was a strong proponent of analogy as a tool of scientific reasoning. The effects of this approach to nature were to be significant in Houghton's studies in the copper district.[13]

Obviously, these changes in the nature and practice of science did not happen overnight or even by conscious decision. Scientists generally, and many geologists in particular, remained explicitly committed to a strongly empirical and observational approach to nature, at least in their public pronouncements. Neither did these changes imply, in any way, that controversy within geology had ceased. Nevertheless, the common ground over which such internal controversies might occur was increasingly being acknowledged. Geology was undergoing a significant and far-reaching change.[14]

And yet, virtually all of those who had made contributions to the knowledge of the district to this point were clearly not professionals in the study of the earth. To the degree that their efforts were indicative of any scientific philosophy at all, it would have undoubtedly been an unconscious empiricism. Little attention was given to, or faith placed in, useless theorizing, and there was little of a consistent body of knowledge against which their observations could be judged or evaluated. They were "democratic fact gatherers," primarily concerned with demonstrating or confirming the relatively simple question of the bare existence of the native copper. Beyond that, they were not inclined to support or deny considerations that would have to have been based largely on theoretical grounds.

After about 1820, however, all of this began to change; the traditional picture of science as an aside of travelers and explorers, or as something done by amateurs in their leisure time, was being challenged by the emergence of the specialists. As one result, the earlier simple question of the *existence* of the native copper was gradually being superseded by the more complex and theoretical question of its *meaning*, and on this question there was to be much divergence of opinion.

Therefore, I will try here to clarify the events of this period by recognizing within them, and among the persons involved, two different perspectives, two distinct "traditions." For convenience, these will be called the "amateur" and the "professional" traditions, each of whose proponents differed in background, purpose, and goals. The first of these, the amateur tradition, extended as far back as the earliest European awareness of the copper in the new world. The first clear evidence of the second, the professional tradition, along with the later divergence of the two, is found in these early decades of the nineteenth century as part of the professionalization occurring within science. Until midcentury these traditions operated as two parallel streams in the interpretation of the significance of the native copper.

The amateur tradition was usually based on some direct, personal observation and examination done within the district itself. Its advocates, well educated or not, were strongly empirical and observational in outlook, and were in some sense "public" figures. Although some of them were well read, they were not geologists or experienced miners, nor did they represent any geological school of thought. Scientifically, they were amateurs, and many of them were simply travelers or explorers. There was a ready and waiting market for any news from the newly opening west, and such amateurs often disseminated their observations and conclusions in writings intended for public consumption. They tended to be noncritical in their approach to the meaning of the native copper, accepting what they found at face value, without much concern about whether those findings conformed to any preconceived notions.

The professional tradition, in contrast, was maintained by those who, though perhaps at first educated in some other field, were at least in some sense scientists, members of a group united by a common and growing tradition. These persons, though generally acknowledging or even insisting on the primacy of observation over theory, nevertheless were more inclined to be guided in their interpretations of nature by the totality of past experience. Based on the experience that had been gained in other copper-mining regions, the professionals tended to be much more cautious, or perhaps even negative, in their judgments on the importance of native copper.

Broadly speaking, the amateur tradition, with its traders, explorers, stories, letters, samples, and tales of distant wonders such as the Ontonagon Boulder, dominated through the end of the eighteenth century. In the work of Henry R. Schoolcraft we see most clearly the first signs of the divergence of these two schools of thought and the first clear indication of the tension between them. Aspects of these traditions are also reflected in the work of Houghton, Jackson, and several of the others who were active in developing the region. The two streams were not to become united into a consistent, unified perspective until after the century passed its midpoint. By that time the Keweenaw had become established as one of the major copper-mining districts of the world, and the true nature of its copper and the lodes in which it was found had finally been recognized.

3

Henry Rowe Schoolcraft Meets the Keweenaw

The dawning of the nineteenth century saw the Keweenaw copper district officially become a part of the United States. Remote though it was, no longer would plans and efforts have to span an ocean in the attempt to find fulfillment. The Great Lakes were soon to become well-traveled avenues as the drive to the west became a part of America's sense of national destiny. The federal government began to see the possibility of a domestic source of copper as an increasingly attractive alternative to dependence on foreign sources. Through the first half of the century the pressure for development grew, and it culminated with the opening of the district to full-scale mining at midcentury.

The Keweenaw Becomes American

After the achievement of independence by the United States, the vast region to the west of the original colonies had been claimed by several of the new states, for they wished to extend their influences into this undeveloped area. After some debate, these states reluctantly agreed to give up their claims in favor of those of the federal government. One of the results was that the land lying to the northwest of the Ohio River was incorporated into the Northwest Territory by the Ordinance of 1787.

In the earliest years of the nineteenth century the Northwest Territory was administratively divided in several ways, and in 1805 the Territory of Michigan was established.[1] However, the copper district was not always in this Michigan Territory. It lay within the western portion of the south shore of Lake Superior, an area shifted from one territory to another as administrative efficiency dictated. It became part of the state of Michigan only as an afterthought when statehood was attained in 1837, and then only with much opposition from within Michigan itself.

During and for some time after the turbulent years surrounding the achievement of American independence, the district hovered in the background of national awareness; the great demand for copper by the electric industry still lay in the distant future. Nevertheless, the metal had long been essential to commerce and industry, and its use was already increasing significantly. Pins, nails, pots, pans, wire, boilers, musical instruments, engraving plates—all these were made of the red metal. Protective sheathing for naval vessels was taking ever larger tonnages. The traditional uses of copper in coinage and in brass for scientific instruments and decorative furnishings, and in bronze for statues, bells, and cannon, were increasingly important. Although effective American control over the Great Lakes region was not established until 1796, the possibility of a domestic supply of copper with the resultant freedom from dependence on foreign sources was becoming more and more desirable.

It was evidently in 1800, during the presidency of John Adams, that the first specific attention was given to the copper region by the United States government. In 1798 Britain had placed restrictions on its own exports of copper, and the effects were distinctly felt in the United States. The copper needs for the ships of an expanding navy, along with the limited resources known at the time in the eastern states, led Congress to consider developing possible domestic sources of the metal in the West. The copper on Lake Superior was already known well enough to merit an official investigation by Congress, and a resolution was passed April 16, 1800, that authorized the president

> to employ an agent, who shall be instructed to collect all material information relative to the copper-mines on the south side of Lake Superior; and to ascertain whether the Indian title to such lands as might be required for the use of the United States, in

case they should deem it expedient to work the said mines, be yet subsisting, and, if so, the terms on which the same can be extinguished.[2]

Richard Fenimore Cooper, brother of the author and son of the New York congressman who introduced the resolution, was appointed to be that agent. Preparations were made and an expedition was outfitted, but because of the lateness of the season it was delayed until the next spring. By that time, however, Jefferson had succeeded Adams, and in the altered political climate the expedition was no longer a priority. Notwithstanding the protests by the Coopers about the potential value of the mines, the expedition was canceled before ever leaving New York. The equipment was sold for what value it had, and America's first official attempt to direct a national effort toward the mineral wealth of the far Northwest abruptly ended, with nothing accomplished.[3]

Nevertheless, even though abortive, this initial effort had effects that reached far beyond those immediately apparent. It helped create a climate of awareness that reached both into the halls of government and beyond to the public at large. Through reports, newspapers, and even literary journals the south shore of Lake Superior, and the Ontonagon River in particular, became associated in the public mind with great potential riches.[4] Even at this early date in the nation's history, the seeds were being scattered that were to blossom into the copper rush of the 1840s.

One of those who succumbed early to the copper fever was Francis Le Baron, a surgeon sent by the war office to the military garrison stationed at Michilimackinac, at the strait between Lakes Michigan and Huron.[5] There, at the crossroads of the upper lakes, while caring for the soldiers and Indians, he had repeatedly heard stories of the copper to the west. He had also seen and obtained specimens, and they impressed him greatly. In 1810, prompted by the arrival at the garrison of a store of medical supplies, he responded with a letter to William Eustis, the secretary of war. Although Le Baron was a doctor, his concern was not medicine but copper.

From his long tenure in the area he had become convinced that the copper in its "virgin state" might well constitute an "inexhaustible source of wealth" in the region to the west. He admitted that, in regarding the potential of such mines, he might be "too sanguine in my Ideas of their richness." Yet, he felt that an official investigation was certainly merited, for "that valuable & necessary Metal is growing daily in greater demand, & the power of procuring it is daily more difficult & uncertain." Le Baron had already taken the liberty of sending some of his specimens of native copper to a prominent friend in Boston and to a member of Congress. Finally, in his letter to the secretary, he came to his real point:

> Should you think from what I have said, or from what you have probably heard from Others, that the thing is of sufficient importance to investigate, I should wish that before My departure from this part of the Country to be sent there; early on the opening of navigation next Summer, would be a favorable opportunity as Men could then be procured eisily [sic], & should You conclude to send Me it would be requisite to send a person from the States who is a practical Mineralist. . . . If however You should think the Thing not of sufficient importance for the trouble & expence; or Not inclined from other Motives, I would beg from You permission for leave of absence to accomplish such a Voyage Myself.[6]

Clearly, Le Baron had been bitten hard by the copper bug, and his enthusiasm was evidently contagious. The day after the secretary of war received Le Baron's letter, he began inquiring after the person that Le Baron had requested. He wanted to know "whether there is to be found in Phila[delphia] a man who has worked in mines, of sober habits, a practical mineralist, who would be willing to go to Detroit & from thence to Lake Superior, to accompany a Gentleman who will be selected for the purpose of exploring the Copper mines on the margin of that Lake."[7]

Eustis, in his governmental position, evidently saw the national significance such deposits might have. He went beyond Le Baron in suggesting a trip of six to nine months, though Le Baron had spoken of only three, and also seems to have anticipated the possibility of some actual mining. It seems clear that the dichotomy between the amateur and professional traditions already existed here in this first decade of the nineteenth century. When, in response to Le Baron's request for a "Mineralist," Eustis recommended that the person be "an experienced *Workman*, not a theorist [his emphasis]," this distinction already seems explicit, and his sentiments were obviously not with the theorist. War with the British was looming, however, and the region at Michilimackinac was to be directly involved. It appears that these plans, like those of a decade earlier, never came to fruition.

Even so, the reputation of the copper on Lake Superior continued to spread, and its potential for the future of the country was increasingly noted. Albert Gallatin, secretary of the treasury in 1810, struck a familiar note when he pointed out that the lack of a strong domestic copper industry was because of the duty-free nature of imports and not the lack of sources of the metal, for "rich copper mines are found . . . near Lake Superior; but they are not now wrought."[8] Neither did Le Baron's interest in the copper abate, and he continued to promote the district by taking it upon himself to send additional pieces of copper to governmental officials.

One mass that he sent to Samuel Mitchell of New York about 1817 apparently received much notice in newspapers, and at least one other made it to Europe. Earlier, Le Baron had given some copper samples to Eustis, the secretary of war. In his later role as "Minister Plenipotentiary and Envoy Extraordinary from the United States," Eustis presented one of them to officials in the Netherlands for deposition at the University of Leyden. Eustis requested that an analysis of the piece be made there, both to provide a basis for comparison with analyses done in America and to "add to the diffusion of information respecting our country."

This analysis, done by the assay-master at the mint at Utrecht, resulted in a report published in several American periodicals.[9] It stated that the copper from North America had been fused by "a process of nature," and that the metal was absolutely pure, for the testing "proved that it does not contain the smallest particle of silver, gold, or any other metal." The specimen was "incomparably better than Swedish copper" and was "peculiarly qualified for rolling and forging," not requiring any smelting whatever. Knowledge of Keweenaw copper was rapidly disseminating both at home and in Europe, and increasingly the southern shore of Lake Superior was achieving a widespread reputation.

Most of these preliminary efforts, even though they helped the public focus on the copper region, did little to contribute to any real scientific understanding of the true nature of the deposits or their geological setting. Despite the speculation, no realistic assessment of the economic potential for mining had ever been achieved. The remote locality, combined with the still limited market for the copper, ensured that the pressure for economic development of the region would be small. However, all this was to change significantly as the country continued to develop, and industrialization, slowly at first and then ever more rapidly, soon went hand in hand with the country's fascination with westward expansion.

In 1820 an expedition organized by Michigan Territorial Governor Lewis Cass visited much of the country to the northwest of the lakes and returned with a wealth of new information about the region. The pursuit of geographical knowledge and the uncovering of the secrets of the natural world, however mundane, had attained an almost religious significance in the early nineteenth century, and exploratory expeditions had become integral to that outlook. This large expedition had a variety of objectives, both geographical and political, but a major result was an important step being taken in understanding the significance and potential of the Lake Superior copper district. This expedition also marked a watershed in the attempt to understand the nature of the native copper. It introduced a thirty-year period, from 1820 to 1850, during which more progress was made toward that understanding than in all the preceding years put together.

Lewis Cass had been appointed governor of the Michigan Territory by President Madison in 1813 (fig. 8). Quickly becoming one of the territory's greatest boosters, he set out to create a climate in which the forces of law and order would assure incoming settlers of adequate protection in a frontier environment. Although convinced of the inevitability of the expansion of the country to the west, Cass was also widely known for his fair and benevolent treatment of the Indians. As a champion of their rights, his ample girth earned him their affectionate title of "Big Belly."[10] His plans for the region, however, faced problems, not the least of which was the lingering effect of the war with the British.

The war of 1812 had a disastrous effect on the Michigan Territory, leaving it depleted of people and supplies and its administration in disarray. Cass's hope of enticing settlers from the eastern states was being frustrated by tales of attacks by Tecumseh's braves and stories that Michigan was an "interminable swamp" with virtually no good land for cultivation. Unfortunately, such tales seemed confirmed by opinions being expressed by influential persons. Cass faced the major task of rebuilding confidence in the region as a place with a future, and he began a vigorous propaganda campaign that soon won over most of the skeptics. His efforts culminated with an official visit by President Monroe in 1817, and with

Fig. 8. Lewis Cass, the first territorial governor of Michigan and leader of the expedition of 1820. Courtesy Michigan Historical Collections, Bentley Historical Library, The University of Michigan.

the first public land sale in 1818 the development of the territory was well under way again. Cass had effectively countered the somber rumors, and they were soon replaced by more idyllic tales. The East soon resounded with the cry, "On to Michigania."[11]

President Monroe also had a clear vision of the significance of the West in the future development of the country, and he brought into his administration another who shared that vision, John C. Calhoun. As secretary of war, Calhoun was particularly interested in ensuring an adequate military presence in the West. To establish that presence he had organized military posts in the newly opening regions. Cass was then quick to realize that a federally sponsored expedition to the upper Great Lakes might serve both the military security and the future of the Michigan Territory, enhancing its desirability as a place for settlement.

In 1819 Cass wrote to Calhoun, suggesting that such a trip might have mutually desired beneficial results: "The country upon the southern shore of Lake Superior has been but little explored, and its natural features are imperfectly known. It has occurred to me that a tour through that country, with a view to examine the productions of its animal, vegetable, and mineral kingdom . . . would not be uninteresting in itself, nor useless to the Government."[12] Cass indicated six "political objects" that would be given attention along the proposed route. Four involved relations with the Indian tribes that would be encountered, reflecting Cass's concern for their future welfare. Another concerned the British fur trade in American territory. The remaining one (Cass's third) is of interest here:

> Another important object is the examination of the body of copper in the vicinity of Lake Superior. As early as the year 1800, Mr. Tracy, then a Senator from Connecticut, was dispatched to make a similar examination. He, however, proceeded no farther than Michilimackinac. Since then, several attempts have been made, which have proved abortive. The specimens of virgin copper which have been sent to the seat of Government have been procured by the Indians, or by half-breeds, from a large mass represented to weigh many tons, which has fallen from the brow of a hill. I anticipate no difficulty in reaching the spot, and it may be highly important to the Government to divide this mass, and to transport it to the seaboard for naval purposes.[13]

The mass was the great boulder on the bank of the Ontonagon River. Although he proved a poor prophet of the ease with which it could be reached, Cass here exhibited a clear understanding of the significance of the copper and the need for a coherent federal policy for its eventual exploitation. Cass further pointed out to Calhoun that speculation about the copper by the general public had already begun, and he clearly foresaw the possibility that private individuals might try to strike bargains of their own with the Indians. Feeling that such private deals might complicate the orderly development of the copper deposits, he was determined to see the national interest served by obtaining some sort of agreement with the Indians as quickly as possible. His final petition to Calhoun was to be particularly significant: "Should this object be deemed important, I request that some person acquainted with zoology, botany, and mineralogy may be sent to join me."

Calhoun's response was enthusiastic, and he agreed that Cass's six points should be the basis of the expedition. With minor modifications, he said, Cass could consider them to be instructions directly from his office. He promised that "every encouragement, consistent with the means in my power, will be given by the [War] Department." Calhoun proposed sending along some officers from the topographical engineers and, noting Cass's earlier request, he replied that "a person acquainted with zoology, and botany, and mineralogy, will be selected."[14]

In February 1820 Calhoun wrote to Cass to inform him that he had made his selection for the position: "a gentleman of science and observation, and particularly skilled in mineralogy, has applied to me to be permitted to accompany you on your exploring tour on Lake Superior." Calhoun indicated that his choice would be "highly useful to you, as well as serviceable to the Government and to the promotion of science." He further directed Cass to provide him with "every facility necessary to enable him to obtain a knowledge of the mineralogy of the country, as far as practicable."[15] That gentleman, who was to make important progress toward a realistic assessment of the copper region and a clarification of the nature of its copper, was Henry Rowe Schoolcraft (fig. 9).

Henry Rowe Schoolcraft

Schoolcraft was "a man of industry, ambition, and insatiable curiosity."[16] He was born in 1793 and passed his youth in Albany County, New York. Although he received only "a smattering of education of the grammar school variety," he was apparently a prodigy. He studied while other children played, and his active mind was drawn in many directions. What collegiate studies he had were taken only informally under Fredrick Hall at Middlebury College, who found him a willing student and who evidently felt that Schoolcraft exhibited great promise as a chemist. Particularly intrigued by rocks and minerals, Schoolcraft later wrote of himself that "I had from early youth, cultivated a taste for mineralogy, long indeed it may be said, before I knew that mineralogy was a sci-

Fig. 9. Henry Rowe Schoolcraft, pioneer explorer of the Lake Superior region. Courtesy Burton Historical Collection, Detroit Public Library.

ence."[17] He was living astride the watershed between amateur and professional science.

Although one of his primary goals in life was to be recognized as a major author and literary figure, he was involved in several fields of endeavor. With his father, he came to hold a prominent place in the glass-making industry of early America, but a familiar problem, the importation of cheap glass from abroad, put them out of business. He was soon casting about for new experiences, with the distinct feeling that his destiny might well lay in the new country being opened toward the interior.

In 1818 he left New York, and for much of the remainder of his life he was associated with the developing regions of the expanding west. He achieved perhaps his greatest renown for his role in the discovery of the source of the Mississippi River. In 1822 he was appointed Indian agent in the Lake Superior region and the next year he married Jane Johnston, whose mother was the daughter of a powerful Indian chief. Her insights, in turn, helped provide him with the basis for his researches into Indian folklore.[18] He published very extensively on Indian life and culture, and his writings were the inspiration for Longfellow's "Song of Hiawatha." The field of American ethnology was effectively created by the work of Schoolcraft, although it was not his first choice for his life's work. "It grew up about him and enmeshed him, in spite of his desire and definite plan to be a mineralogist and mining engineer."[19]

His introduction to the American frontier came in 1818 when he traveled through Arkansas and Missouri on a journey that totaled some six thousand miles. There, perhaps prompted at first by his interest in glass-making, he visited the lead-mining regions. On that tour, he made careful observations on virtually every mine in the area and assayed mineral samples in a chemical furnace he constructed. Some of his mineralogical observations appeared the next year as *A View of the Lead Mines of Missouri*, in which he demonstrated a clear grasp of the principles of mining engineering and economic geology.[20]

When he returned to New York in August 1819 with the manuscript of his new book, he was still a virtual unknown. After its publication he sent copies to anyone and everyone who might possibly be able to throw a favor his way, including "deluxe bound, gilt-edged copies" to President Monroe and other governmental officials.[21] As a result, by the time he appeared in Washington that November, he had already begun to acquire a reputation as a mineralogist. The value of publication in putting one's name before the public eye, and the hunger of the public for information about the opening of the Mississippi valley region, were lessons he never forgot.

Despite his reputation, however, Schoolcraft's college education had been almost entirely informal, a fact about which he was sensitive for the remainder of his life. His mineralogical work was always largely empirical and observational, and the many publications that grew out of his expeditions and travels were intended primarily for public consumption. Here, at the beginning of the "emergent period" of professionalization in geology, and despite the expanded reputation he was to achieve in the future, he was largely a representative of the amateur, public tradition in his approach to the natural world.

His book on the lead mines brought him to the attention of Calhoun and that, in turn, resulted in the offer to accompany the Cass expedition. Schoolcraft was already familiar with the rumors about the copper. Stories about the famed Ontonagon Boulder were making their way through the lead mining camps of Missouri, and while there he had already tried unsuccessfully to have himself appointed to lead an expedition to the copper region. He had also included a reference to the mineral wealth of Lake Superior in one of the many poems he was so inclined to write.[22] Now, although he regarded the pay as low, he recognized the position on the Cass expedition as "the bottom step in a ladder which I ought to climb."[23] His acceptance of the appointment began his long association with the Great Lakes area, and also resulted in his having a major part in developing a scientific picture of the copper deposits of the Keweenaw.

With Calhoun's approval to proceed, Cass enthusiastically began organizing the expedition by ordering large birchbark canoes from his friends among the Saginaw Indians. He busied himself with soliciting French voyageurs on the streets of Detroit, selecting books and scientific instruments, and choosing bright associates to fill positions on the journey. Capt. David Douglass was engaged as an engineer to perform many of the scientific studies not assigned to Schoolcraft. Cass again emphasized the significance of the copper to Schoolcraft, stating that "a complete examination of this copper is an important object."[24] The long preparations finally being completed, on May 24, 1820, forty explorers, soldiers, voyageurs, and Indians departed Detroit for the North in three large birchbark canoes.

The Expedition of 1820

The expedition began by proceeding up the west coast of Lake Huron to Mackinac Island, where additional supplies and more soldiers were picked up. In mid-June Cass presided over a major conference with the hostile Indians at Sault Ste. Marie, whose loyalties were still clearly with the British. After some tense moments, during which Cass strode alone into the Indian camp, confronted a chief dressed in a British uniform, and ground a British flag under his boots, agreement was reached. The region around the rapids was ceded to the United States, thereby making a permanent military and governmental presence possible. The soldiers returned to Mackinac, and the expedition prepared to proceed westward along the south shore of Lake Superior toward their next major objective, the copper rock on the Ontonagon River (fig. 10).

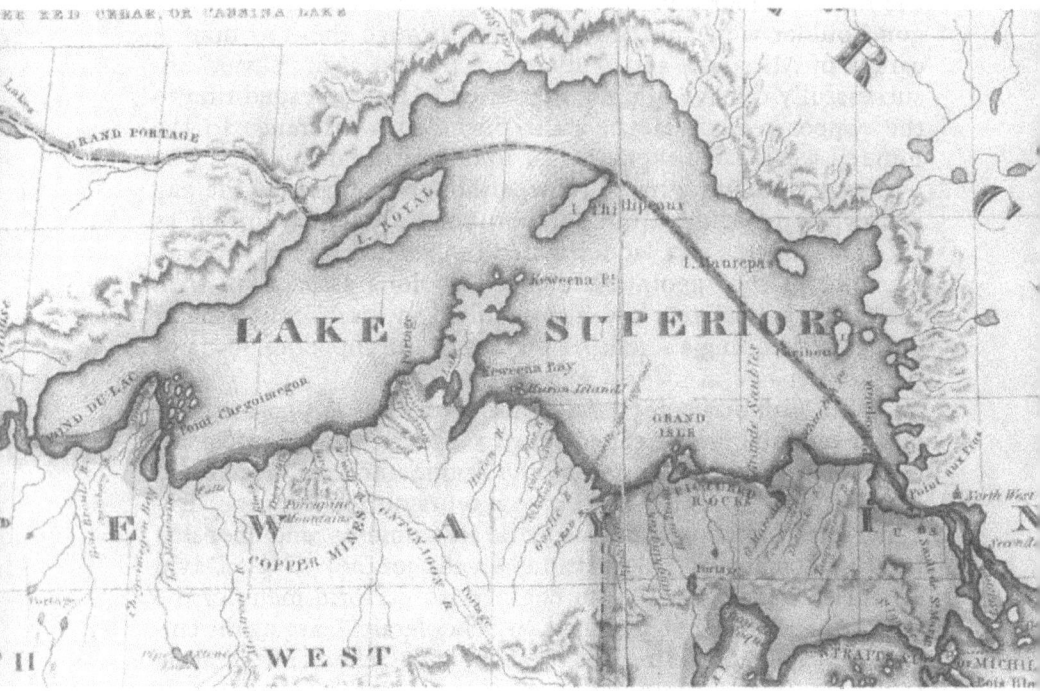

Fig. 10. Route of the Cass expedition of 1820, as published in Schoolcraft's Narrative Journal of Travels *of 1821. The dark line along the south shore is a contemporary shading of an underlying dotted line showing their route. Copper mines are indicated at the site of the Ontonagon Boulder.*

As they pressed along the shore the group marveled at the sights, especially the vast dunes of the Grand Sable Banks and the steep sandstone cliffs of the Pictured Rocks, which Cass called "two of the most sublime natural objects in the United States."[25] Schoolcraft regretted not being able to stop as frequently as he would have liked to study the geology, but specimens were collected as regularly as possible as they went along. The cliffs of the Pictured Rocks, in particular, brought out Schoolcraft's characteristic tendency to mingle scientific observations and flowery, rhapsodic prose or even poetry as he enjoyed the great works of nature.[26]

Pressing onward they soon passed the spot, just east of the present city of Marquette, where the ancient, primary rocks of the western portion of the Upper Peninsula appeared along the shoreline. Schoolcraft made note of the change, and included many astute geological observations in his running account. A week of constant paddling, including a harrowing crossing of Keweenaw Bay in very rough weather, brought them to the Keweenaw peninsula and the beginning of the country where copper was most frequently found. There, rather than lengthen the voyage excessively, they took the traditional portage route by ascending the Portage River. After paddling the length of Portage Lake, a short overland passage brought them back to Lake Superior on the northwestern coast of the peninsula. In so doing, they did not make the trip around the tip of Keweenaw Point, and they therefore did not see the famous "green rock" that outcropped along the shore.

After the portage, a headwind temporarily prevented them from continuing on their journey southwestward toward the Ontonagon River. They therefore contented themselves for a time with searching the pebble strewn beach for items of interest. Schoolcraft discovered a mass of native copper of nearly two pounds, along with many other rocks through which fine copper grains were disseminated. In addition, minerals of many varieties were found, so much so that several of the voyageurs and soldiers, "who had before felt no interest in this study . . . also turned collectors."

Schoolcraft was particularly fascinated to find that his Indian guides were also anxious to search, and that they found many of the best specimens. They repeatedly expressed their desire to know the reasons for his interest, and he tried, with great difficulty, to come up with appropriate Chippewa words that would convey to them something of what he was doing and why. The Indians, in turn, bestowed on him the name Paw-gwa-be-can-e-ga, "the destroyer of rocks," which, he soberly noted, "may be considered as synonymous with the word 'Mineralogist.'"[27] Then, the weather having settled, they rose at four the next morning and, as Charles Trowbridge, one of the expedition members put it, "with a fair wind embarked and steered for the River Ontonagon, one of the moving springs of our expeditions [sic], for here is deposited the enormous mass of copper so much spoken of in the civilized world."[28]

On the twenty-seventh of June 1820, after a journey of ten days from the Sault including layovers, they arrived at the mouth of the fabled river with its Indian village on the west bank. They now had the opportunity to examine the old stories and rumors in a clear light, "to ascertain how far these accounts are founded in truth."[29] Cass was determined to lose no time in getting to the task at hand. Leaving most of the party at the mouth of the river, which keenly disappointed some, they started for the interior. Schoolcraft was particularly conscious of the significance of the occasion. He and Cass, along with several other principal members of the expedition, voyageurs, and four Indian guides engaged at the Chippewa village, enthusiastically proceeded in two light canoes. Up the river they went, facing a journey of over twenty miles, bound at last for the "so much wished for pleasure of seeing that wonder of the North western world."[30]

After camping overnight, the party split up. Schoolcraft's group continued overland along an old Indian path toward the copper, while Cass's party went along the river, which required frequent portages around the many rapids. The parties met to finish the journey together, but alas, Cass could not continue. Clambering over hills and around portages proved to be a little too much for the chief executive. The heat of the day, "90° under the dark shade," combined, no doubt, with the attributes that had earned him his Indian title of "Big Belly" took their toll, and Cass found it necessary to allow the party to complete the journey to the boulder without him.[31] To take him back to the canoes, he was given an Indian guide who then departed "with the Executive of the Territory, wearied and perplexed, at his heels." The two of them, in turn, got lost in the woods and wandered for hours; Cass became totally exhausted, and they were reunited with the group returning from the boulder only late that night.[32]

Finally, after great exertion by those remaining, their goal was reached. The next year Schoolcraft published a sketch of their arrival (fig. 11), but it shows little regard for accuracy. Three canoes are shown approaching an immense boulder, but in fact their *two* canoes had been left below the rapids and they had hiked the last six miles over rough terrain. They found the mass lying on the bank of the river at the foot of a high clay bluff from which, Schoolcraft inferred, it must have fallen at some remote time. Perhaps the extreme effort of the trek contributed to their disillusionment, but their first feelings were of great disappointment at its size. Douglass, who called it "a mere stone, a large pebble," wondered if perhaps it was not the mass they were seeking, although the Indian guides assured him that it was.[33] It was only about four feet

Fig. 11. The Cass expedition at the Ontonagon Boulder, as drawn by Henry Schoolcraft, based on his visit of 1820 and as published in his Narrative Journal of Travels of 1821. Courtesy Michigan Technological University Archives and Copper Country Historical Collections, Michigan Technological University.

across, and Schoolcraft estimated the weight of the copper in it at twenty-two hundred pounds; they had been expecting something much larger.

Gradually, however, Schoolcraft warmed up to what he was seeing. The mass was brilliantly metallic where the river had scoured it at high water, and he soon conceded that it "is nevertheless, a remarkable mass of copper, and well worthy a visit from the traveler who is passing through [!] the region." Looking over the wild scene from the top of the bluff, he wrote that "one cannot help fancying that he has gone to the ends of the earth, and beyond the boundaries appointed for the residence of man," a remark that, when published, earned him the derision of some of the other members of the group.

After making many observations on the copper, some small specimens were obtained with much difficulty, despite the special tools that had been brought for the occasion. Schoolcraft recognized the likelihood that "extensive mines of this metal exist in the vicinity," and bemoaned the lack of time to search for the metal within the rock itself.[34] By evening, the party reunited and bedding down for the night, he was already reflecting on the events just past. Poetic allusions once again came to mind, and later he wrote, "But deeply printed on our memory, and long to remain there, are the thrilling scenes of that day and that night."[35]

Poetic allusions were clearly not in everyone's mind. Charles Trowbridge, the assistant topographer who did not accompany the group to the boulder, recorded that the party returned to the river's mouth the next day "fatigued beyond description and disappointed in their expectations." After hearing Douglass's disparaging account of the boulder, he lamented, "Thus ended all the marvellous stories we have heard about the copper mines of Lake Superior, which some have gone so far as to represent [as] inexhaustible."

The entire expedition prepared to move on, but head winds prevented their departure and they spent the next day on the shore. That evening, in the village at the mouth of the river, the Indians entertained them with dances that Schoolcraft, the future Indian agent and expert on all aspects of Indian life, found to be a severe tax on his patience. The ceremonies included the presentation of a silver medal to, oddly, the Indian guide who had led Cass astray in the woods, but perhaps he had compensated by making an impressive speech professing his loyalty to the Americans. A nine-pound mass of copper that an Indian had found in the riverbank was also given to Schoolcraft before he left. The next morning, July first, conditions were favorable, and the expedition pushed on to the west, a major objective of their journey having been achieved, although Trowbridge was "well pleased to leave a place where we had met with so much disappointment and fatigue."[36]

The remainder of the expedition was a great success. They continued west along the south shore of Superior to the head of the lake, then up the St. Louis River to Sandy Lake. There, a quick venture up the Mississippi River was made to its "source" in the lake then named for Cass. In Schoolcraft's later writings the search for the source of the Mississippi is represented as the purpose of the expedition, even though it was not one of the initially stated objectives. They went back down the Mississippi, had more conferences with the Indian tribes, went through what is now Wisconsin to Green Bay, and then arrived at Lake Michigan. There they split up, later to regroup back at the Straits of Mackinac. Finally, after having traveled some four thousand miles in 122 days, they reached Detroit on September 23, their mission well accomplished.

The great number of communications and publications that grew out of this expedition influenced and molded the views of the nation strongly on the subject of the upper lakes region in general and on Keweenaw copper in particular. Shortly after returning to Detroit, Cass sent Calhoun a preliminary report in which, as part of his overview, he summarized their observations on the copper. Perhaps forgetting momentarily, in his enthusiasm, that he had not made it to the boulder, he described how "we ascended that stream [the Ontonagon] about thirty miles to the great mass" and how "our chissels [sic] broke like glass." He also expressed his intention to send Indians the next spring to procure larger specimens from the mass, which, he said, he would send along to Washington for distribution to "the various literary institutions of our Country." The major report on the copper, he commented, would soon come from Mr. Schoolcraft, that being the object "to which his enquiries were directed."[37]

The Schoolcraft Reports

Schoolcraft was well aware of the weight his views on the copper district would carry. Less than two months after his return he submitted his specific observations on the copper to Calhoun in a long letter that constituted his official report. In that report, dated November 6, 1820, Schoolcraft reviewed the earlier efforts of those who had addressed the copper deposits between 1689 and the beginning of the Cass expedition.[38] He pointed out that masses of copper from the south shore of the lake had long been known, that such masses had found their way to different parts of the country and to Europe, and that some of these had already been acquired by the government.

A special point was made of the mass that had been found on an island off Point Chegoimegon in what is now Wisconsin. That was the piece from which the war department had obtained a specimen and of which an analysis had been made at the University of Leyden. The results of that analysis, which had been made in 1818 and which had demonstrated that the material was ". . . metallic copper in a state of uncommon purity . . . ," had been reported in many periodicals. All these

> have operated either in producing public belief in the existence of copper mines on lake [sic] Superior; or in conferring public notoriety upon their extent, importance, and location; while a general agreement has been observable that the southern shore of the lake is most Metalliferous, and that the Ontonagon river may be considered as the seat of the principal mines. . . . It has been reiterated in several of our literary journals, and in the numerous ephemeral productions of the times, until the public expectation has been raised in regard to them.

In describing the Ontonagon boulder, he acknowledged its diminished size relative to most of the old stories, but "notwithstanding this reduction, it may still be considered one of the largest and most remarkable bodies of native copper upon the globe." The mass, he noted, was not totally pure copper, but was intermixed and interlaced with a dark rock, a condition noted in the journals of others in the party also.[39] This rock he identified as serpentine, which, because it was not found near the mass, indicated to him that the mass must have been transported by some agency from a great distance. This conclusion would therefore apply with equal force to all the masses, large or small, that have been found near the surface anywhere in the district.

Was the copper in amounts great enough to mine? If so, what kind of ores of copper would likely be found? Would they be copper compounds as in most other districts? Would it be native copper? Both, perhaps? It is clear that Schoolcraft considered questions such as these important, and he pondered them for some time. His views, however, as he expressed them over the years in letters and publications, are not entirely consistent.

Fortunately, he was firmly convinced of the value of publication. In the years after the conclusion of the expedition, he produced several additional reports and sometimes republished the same one several times. Occasionally, similar accounts appear with subtle variations, in the light of which it is possible to sense something of the interplay between observation and theory, between the amateur and professional traditions, that engaged his mind as he addressed this important question of the nature of the ores.

The official report mentioned above, of November 6, 1820, exists in at least four distinct forms, the first being the letter to Calhoun. The next year, on February 16, 1821, the report was submitted to Silliman's *American Journal of Science*, where it appeared with slight but significant modifications from the original.[40] This journal was, by any standard, the most important scientific periodical in the United States. Silliman was, at this time, considered a "champion of the professional interests," and his journal concentrated particularly on geological and mineralogical themes.[41] He had solicited the publication of the report, and Schoolcraft, always aware of the value of wide exposure, readily agreed. He did nevertheless note that "I have probably said more on the geology of the country than may be thought important to the statesman, and less than will be considered satisfactory to the professed geologist and scientific amateur." The report then appeared for the third time as an appendix to another report he submitted to the Senate in 1822. Finally, many years after the events it described, it was included in his *Summary Narrative* of 1855, again as an appendix and again with slight variations from the earlier versions.

Besides this report, early in 1821 Schoolcraft's journal of the trip, along with many scientific observations of the expedition, appeared in book form, although not without some controversy; one member of the expedition called it "a work of *imagination*."[42] In October of 1822 Schoolcraft submitted another report directly to the Senate in response to its request for further information on the copper mines. This report appeared again, with slight modifications, as another appendix in the *Narrative* of 1855. Then, in July 1823, he submitted to Calhoun, and to Silliman for publication, even more remarks on a "recently discovered copper mine on Lake Superior." Through all these reports, public documents, and letters, Schoolcraft laid the basis for what became the national understanding of the region for over a decade. They make it possible to reconstruct, to at least some degree, the factors that influenced his thinking on the nature of the ores.

From the days of his youth Schoolcraft's interest in science had always been distinctly inclined toward the practical as opposed to the theoretical. His early experiences with glassmaking, his detailed study of the lead mines, his chemical analyses of the ores, his skill in draftsmanship—all were a part of the emphasis on the tangible that seemed essential to his personality and which continued to be in evidence in his writings throughout his life. This tendency to emphasize the pragmatic and, in turn, to be suspicious of abstract reasoning, to emphasize observations and facts as opposed to theories and speculations, was widespread at the time. It was a characteristic essential to American science generally up to the turn of the century.[43]

Such an approach to nature was perhaps to some degree inevitable and even appropriate in anyone involved in an expanding frontier environment, where many new and not always immediately understood findings were frequently encountered. It also was consistent with the largely self-taught nature of Schoolcraft's education and his lack of formal training in the sciences. In an article published in the spring of 1820, just before leaving New York to join the Cass expedition for the journey to the north country, he clearly expressed his feelings on the appropriate relationship between theory and fact: "Our theories should be the result of observation, and facts should never be distorted by theory."[44] He therefore exhibited many of the characteristics of the amateur scientific tradition.

Even before he wrote his official report to Calhoun, Schoolcraft had already begun to make some public statements on the copper district that had grown out of his observations on the trip, and one of these is of some significance. In October 1820, just a month after returning to Detroit, he wrote a long letter to Nathaniel Carter, the editor of an Albany newspaper, in which he outlined the expedition and its results. Parts of the letter appeared in that paper, including the following:

> The evidences of the existence of copper in the basin of Lake Superior are ample. There is every indication of its abundance that the geologist could wish. Nature has here operated on a grand scale. By means of volcanic fires, she has infused into the trap rocks veins of melted metal, which not inaptly represent the arteries of the human system; for wherever the broken-down shores of this lake are examined, they disclose, not the sulphurets [sulfides] and carbonates of this ore, but fragments and lumps of virgin veins. These the winds and waves have scattered far and wide.[45]

These comments are important for several reasons. This letter preceded his official report and was written somewhat informally, as a letter to a friend. Presumably, then, it can therefore be taken as representing Schoolcraft's initial, unbiased opinions, based on his views of what he had observed and inferred, without concern about whether they corresponded to any generally held opinions of the time. Although he did not see metallic copper within the bedrock, he here explicitly postulated that the loose masses were originally part of veins of pure metal. He further inferred that these postulated veins were in the traprock, from which the fragments of the metal had been liberated by erosional processes. Here was a clear accounting for their long-noted presence in the surficial deposits of the region, and one that was noticeably more precise than the theories of the previous century.

Even more interesting, however, are his brief but succinct comments on the nature of the ores. Schoolcraft here, very clearly, seemed to take a major step toward providing an answer to that fundamental question. He concluded not only that the native copper was abundant, not only that it had originally come from veins in the bedrock, but he also drew the explicit conclusion that ores of sulfide and carbonate compounds were *not* present in significant amounts. Taken simply at its face value, this statement of Schoolcraft's could well be construed as a remarkably farsighted and accurate brief analysis of the mineralogical situation as it was later revealed to be. Significantly, these views appear to have been based entirely on his own observations. He does not express any concern that these views may be contrary to anything he knew about the mineral deposits, and he wrote those views knowing they were to be published in a newspaper and would therefore become part of the public record.

However, he also was required to write a more formal report for the government, and he apparently realized that this would mean that his views would become, in some sense, the "official story." This evidently raised some concern in his mind over his lack of formal background in mineralogy. Schoolcraft was always sensitive to his lack of formal education, even to the extent that in his later years he embellished his past with "imaginary college attendance."[46] Therefore, to prepare for writing this official report, he apparently decided to supplement his background by reading what books he could find on mineralogy specifically. These would then provide him, the public figure, with a more professional perspective, and would fill in his background on the meaning and implications of native copper in a mining district.

What he found was not what he had expected. He discovered that his straightforward analysis of the significance of the native copper, based on his simple "fact-gathering" approach to nature, was contrary to the consensus of the experts, who did not share his enthusiasm for copper in its pure form. As he studied the opinions of the times, his own interpretations began to shift subtly to accommodate what he was reading. As a result, when he submitted the report to Calhoun less than two weeks after writing the letter to Carter, it contained qualifications and disclaimers that had not been in the earlier letter, and which now modified his simple and straightforward earlier conclusions.

He began the Calhoun report by first clearly stating his original conclusion that the loose masses of native copper were a favorable evidence of larger bodies of the metal nearby in the earth. Curiously, he then almost apologized for drawing that conclusion, and in trying to justify himself he implicitly acknowledged the amateur-professional dichotomy: "Perhaps there is no one, either learned or illiterate, with all the circumstances of a recent discovery placed strongly before him, who would not place a certain degree of reliance upon the indication." He believed, however, that he now had to give the experts their due, even though their ideas went counter to his original conclusions:

> [The presence of native copper] is not, however, an unerring guide, and appears to be liable to many exceptions. A detached mass of copper is sometimes found at a distance from any body of the same metal, or its ores; and these, on the contrary, are often found in the earth, where there has been no discovery of metallic copper to indicate it. So far as the opinions of the mineralogical writers can be collected on this point, they teach:
> —that large bodies of native copper are seldom found together but frequently disseminated among the rocks, spares [spars] and ores of mines; and when found upon the surface, are rather to be considered as tokens of the existence of the Sulphuret [sulfides]
> —the carbonate, and other ores of copper within the circle of country where they occur, than as the precursors to contiguous bodies of the same metal.[47]

Although he did not identify here exactly who the "mineralogical writers" were whose books he had read, it is clear that Schoolcraft's original thinking, based on what he had seen, had been seriously challenged. His earlier optimism about the native copper did not now carry the same conviction that it had carried just a few weeks earlier. The view that native copper was not regarded as valuable in its own right but was only a "token" of other combined ores was new to him. He now realized that his initial convictions were, to a considerable extent, in conflict with the general teachings of the experts of the day.

With Calhoun's permission, just over two months later he submitted the same report to Benjamin Silliman for publication in the *American Journal of Science*, which would, he knew, give it widespread exposure.[48] He therefore took the opportunity to make some additional minor changes, and these clearly indicate that Schoolcraft was continuing to wrestle with the ore question. He now footnoted several of the mineralogical sources he had consulted, including Parker Cleaveland's *Treatise on Mineralogy*, which was the basic mineralogical text of the day. In justifying himself in now minimizing the importance of the native copper, but directly contrary to his original assertion, Schoolcraft quoted Cleaveland directly: "Native copper is not rare, nor is it found in sufficient quantity to be explored by itself. It sometimes occurs in loose, insolated [isolated] masses of considerable size."[49]

He also added a new section discussing Cleaveland's view, which was clearly the conventional view of the day, that native copper was significant only to the degree that it accompanied the sulfide and oxide ores, as it did in other parts of the world. Schoolcraft's statement in the original report, quoted above, was that the experts teach "that large *bodies* of native copper are seldom found." He now altered that to read "that large *veins* of native copper are seldom found [my emphasis]." Recall that in the letter to Carter he not only had declared that such veins *were* present in the district, but he had compared them to the arteries of the human body.

These readings of the opinions of the experts, then, done only after the expedition had been completed, had a noticeable effect on Schoolcraft and distinctly altered his thinking. He was clearly caught between the two poles of the amateur and the newly emerging professional traditions. On the one hand there was the evidence he had seen and on which he had first expressed himself clearly. On the other hand were the opinions of the mineralogical experts, as expressed in the works he had read and which he was reluctant to oppose. After contemplating both, he drew two conclusions. They exhibit the tension he felt between what he had seen and what he had read, and they also seem to indicate that he was not totally satisfied with them.

First, he concluded that the region

> contains very frequent, and some most extraordinary imbedded masses of native copper; but that no body of it, which is sufficiently extensive to become the object of profitable mining operations, is to be found at any particular place. This conclusion is supported by the facts already adduced, and *so far as theoretical aids can be relied upon*, by an application of those facts to the theories of mining [my emphasis].

Schoolcraft here seems to have reluctantly decided to go along with the thinking of the experts, but he also very astutely recognized that he was now intermixing observation, which he favored, and "theoretical aids," in which he had less confidence.

Second, he concluded that a close examination of the region "would result in the discovery of very valuable mines of *the sulphuret, the carbonate and other profitable ores of copper*; in the working of which the ordinary advantages of mining would be greatly enhanced by occasional masses and veins of native metal [my emphasis]." When it is recalled that less than three months earlier he had explicitly stated, based on his own observations, that such ores were *not* present, it is possible to directly sense the forces that were now pulling him in opposite directions. Schoolcraft was clearly involved in an internal quandary over the appropriate relationship between the native copper and the ores of copper compounds. In his mind, we see here the entire historical question of the uniqueness of the Keweenaw brought to a focus. His own observations and the amateur tradition were pulling him in one direction, and theory and the professional tradition were pulling in the other. And yet, even so, Schoolcraft would not yield his opinions entirely. He still declared that the native copper, even if not the major ore, would nevertheless "greatly enhance" the profitability of any mining venture.

Finally, he concluded, "These deductions embrace all I have to submit on the mineral geography of the country, so far as regards the copper mines."[50] He then closed with some general remarks on the nature of the country as it might affect mining attempts: it was remote, cold, and "presented few enticements to the agriculturist," so that commercial development might still be some distance in the future.

There is little question that this report, despite its intriguing tentativeness, represented a genuine step forward in the understanding of the district. For the first time the district had been examined by someone who not only was capable in what might be called "observational mineralogy" but who could also get the most out of what he had seen. Simultaneously the report also demonstrates the power that a view of nature may have in molding the interpretations of even a capable observer, if that view is perceived to be the opinion of experts in positions of authority. Schoolcraft became clearly aware of current thinking only after having made observations and, to some degree, having already drawn conclusions. Nevertheless he did defer to authority, perhaps partly out of his recognition of the informality of much of his mineralogical training. However, he also did recognize the "theoretical aids" for what they were, and he could not avoid maintaining his original conviction that the native copper would still be essential to the mining future of the region.

These reports went a long way toward representing the potential of the district to the rest of the country, and it is clear that this potential was widely appreciated, particularly in Congress. In May 1822 the Senate again passed a resolution requesting the president to submit any information, from whatever source, on the number, value, and position of the mines on Lake Superior, together with the practicality of obtaining title to the lands from the Indians.[51] President Monroe referred the matter to Calhoun, who referred it to Schoolcraft, as being the one most likely to be able to give a definitive answer. Schoolcraft, now the Indian agent at Sault Ste. Marie, responded, on October 1, 1822, with a new report, one made directly to the Senate.[52]

This report, in some ways, went beyond his earlier one to Calhoun, which was included as the first of eight appendixes (thereby marking its third appearance). He noted that, in the two years since the original report had appeared, the district had attracted the notice of several mineralogical works published in Europe, and that he was continuing to receive copper masses from the Indians (citing all the new locations in another appendix). The lapse of the two years had only strengthened and confirmed his own earlier conclusions. He also included the report from Utrecht that opined that the copper was "incomparably better than Swedish copper," a remark undoubtedly not lost on the senators.

As to the possibility of obtaining title from the Indians, a subject that Schoolcraft, as Indian agent, would have been intimately familiar with, he saw no difficulty whatsoever: "The goods and annuities which they would receive in exchange must be vastly more important to them than any game which these mineral lands now afford." Schoolcraft then turned to the final question of the Senate resolution: "The probable advantage which may result to the Republic from the acquisition and working of these mines."

He first pointedly declined to address any issues of "abstract policy," and instead confined himself to the questions of "the fertility of these mineral beds, and their position with respect to a market—points upon which their value to the nation must ultimately turn." On the first, "the fertility of the beds," he once again took a very strong stand in favor of the native copper. He said that, after observation, reflection, and inquiry

> the result has led me to form a favorable estimate of their value and importance. It is not only certain that a prodigious number of masses of metallic copper are found along the borders of the lake, but every appearance authorizes a conclusion that they are only the indications of near and continuous veins. Some of these masses are of unexampled size, and all present metallic copper in a state of great purity and fineness. . . . The metal is disseminated throughout a much greater extent of the country, and in masses of every possible form and size. Until my facts and data can, therefore, be proved fallacious, I must be permitted to consider these mines not only fertile in native copper and its congenerous species, but unparalleled in extent, and to recommend them as such to the notice of the government.[53]

Here it appears very much as if Schoolcraft had returned once again, despite other authors' opinions to the contrary, to his original conviction that the native copper was the most important copper mineral. There was no mention here of "sulphurets" or "carbonates," but only of metallic native copper, with just the vague phrase "congenerous species" to interrupt the flow of his case. Also, after earlier arguing for, and then renouncing, his belief in the existence of veins of the pure metal, their presence was here once again forcefully asserted. His confidence in his own ability to interpret what he had seen had clearly returned, and he now threw out the challenge that his view must be seriously considered until it could be demonstrated that he was in error. The importance of the opinions of the experts of the day was receding in his mind. He once again began to move away from the "professional" view that native copper was of no importance, and back to his own original view of a district whose future would be based substantially on the mining of native copper.

On the second question, "their position with respect to a market," Schoolcraft made an elaborate case that must have had an impact on the congressmen to whom it was addressed. In his report of two years earlier he had dismissed the subject briefly, suggesting that development lay in the distant future. In the interval, he had obviously done some serious investigating, and he now presented a case replete with appendixes full of figures and tables. He concluded that the admittedly remote location of the district, far from potential markets, was not in itself a serious difficulty. As is true for other metallic mines in all parts of the world, he wrote, "the great fertility of the mines counterbalances the disadvantages of its remote position from the market," and he illustrated his point with examples drawn from all parts of the globe. "Neither the remote position, therefore, of the Lake Superior copper mines, nor the want of a surrounding population, present objections of that force which would at first seem to exist."

An extremely detailed analysis of the costs of transportation from the Ontonagon River to the harbor of New York then followed, a distance he calculated as 1,447 miles. "Let it be recalled that there are no mines of copper [anywhere in the world] situated upon the margin of the sea, and that every quintal of sheet copper, bolts, nails, &c, which we receive from Great Britain, Russia, Sweden, or Japan, is transported a greater or less distance on turnpikes or canals, before it reaches the place of shipment."[54]

Finally, and to those in Washington undoubtedly the most interesting, was the question of revenues and the balance of payments:

> I have no data at hand to show the amount of these [copper] articles consumed in the United States, and for which we are annually transmitting immense sums to enrich foreign States. But those who best appreciate the advantages of commerce will readily supply the estimate. It would be an interesting inquiry to ascertain how much of the sums yearly paid for sheathing copper, bolts, nails, engraver's plates, &c, is contributed to the wealth of the respective foreign States who possess mines of this metal."[55]

The mines at Cornwall were the most valuable in the world, he pointed out, and had produced, after deducting smelting charges, an income of nearly one million pounds sterling in the single year 1810. Schoolcraft had, in effect, made a detailed case for the potential value of the mines of the Keweenaw and had written a blueprint for their place in the future of the nation.

It appears, then, that Schoolcraft had confidently returned to his initial opinions on the value of the native copper over the possibility of mining other ores of copper compounds; his original convictions had apparently overcome the effects of the "mineralogical writers." And yet, there was to be one further complication. Nature had another twist in store for him, and it was to again add an element of uncertainty to his thinking.

In July 1823 Schoolcraft was once more in contact with Calhoun through Cass, this time to announce the discovery of a new possible mining location on the Keweenaw and to pass along some accompanying ore samples. The discovery proved to be significantly different from the native masses that had become so common and were continuing to be brought to Schoolcraft at the Sault by his Indian contacts. This sample was a genuine ore of copper compounds, and it came from a locality near the extreme northern end of Keweenaw Point. Although new to Schoolcraft, the ores found at this location had already acquired a considerable reputation among the voyageurs who regularly traveled the coastline of the lake.

In passing along the south shore of Lake Superior, the Cass expedition of 1820 had made its way across the center of the Keweenaw peninsula, taking the traditional route through Portage Lake. A short portage across a narrow stretch of land allowed the group to reach Lake Superior again on the peninsula's northwestern coast, thereby avoiding the much longer journey around the end of the point. However, this also meant that they did not see a significant part of the shoreline of the district. Voyageurs with substantial loads, however, had always found it necessary to take the long lake route around the entire peninsula.

Near the tip of the point, on the east cape of a natural harbor at the present town of Copper Harbor, a vein outcropped along the shore that was several feet in width and which was made up of copper-bearing minerals, all of which contained copper combined with other elements. This vein, long known as "la roche verte," the green rock, extended in a straight line from the shore toward the interior. It also extended out into the lake as a broad green stripe visible on the bottom, and it was this location to which Schoolcraft's attention had now been drawn.[56]

Hearing of the vein, he sent someone to visit the location and get samples for him to examine. Schoolcraft wrote that "I have taken some pains and been at some expence [sic] to investigate this subject, and to procure specimens of the ore. From all I can learn, the vein is a very large and rich one."[57] Of interest however, is what he found when he inspected those samples:

> It appears from external characters, to be the compact Malachite of authors [checking the experts again] which is stated generally to yield, at the mines of Cornwall and Saxony, from fifty-six to seventy per cent, of oxide of copper, the remainder being chiefly carbonic acid and water. It is consequently among the number of the ores of this metal that are most profitably wrought.[58]

He further expressed the opinion that it would prove to be comparable to the ores "at the best European mines." He closed his letter to Calhoun by enthusiastically describing a new forty-two-pound mass of native copper that an Indian had brought him from the Ontonagon, which "contains small points of native silver!" but it is clear that the green rock was a new factor for him in trying to understand the district.

This discovery, coming just after he had reemphasized his original conviction of the great value of the native copper for mining, threw a new and complicating light on the entire question of the nature of the ores. This represented the first time that a sample of an ore composed of copper combined with other elements, rather than native copper, had been brought to the attention of someone of the stature of Schoolcraft. It was truly a significant finding. Further, it was entirely consistent with, and could well be construed as an evidence of, what the conventional geological wisdom of the day had been saying all along *must* be present in the district. This raised anew the question whether the native copper, abundant though it was, was only the minor frosting on a cake made up primarily of ores of copper compounds, just as in other mining regions of the world.

After having so laboriously formulated his previous insights, Schoolcraft was now clearly impressed with the new finding. Once more he found his mental pendulum swinging to the other side as the possibility of significant quantities of copper combined with other elements reappeared; once again he faced fundamental uncertainty on the question of the nature of the ores. This location proved to be of significance in the future. It was, in fact, where the first ore in the entire district was mined, and the minerals were indeed composed of copper compounds. From an economic point of view, however, this ore had absolutely no significance because it was present in trivial amounts. This Schoolcraft could not have known at the time, and until this became clear, the green rock was an obscuring and complicating factor in trying to formulate a clear understanding of the true nature of the mineralogy of the district.

Schoolcraft's position, then, in the development of the district is important, but not without some ambiguity. Aside from any simple analysis of "right" or "wrong," his efforts to understand the ores reveal an intriguing interaction between the observations he made and the theories of his day. There would seem to be little doubt that, at heart, he always believed that the native copper would be essential to the future mining efforts of the district. It also seems evident that he reached this conclusion almost entirely on the basis of a few observations made over a remarkably short time. He further seems to have always clearly recognized the distinction between information based on direct observation, which he favored, and that based on the conventional wisdom of the authorities of the day, even when that wisdom led him to modify his views.[59]

American Responses

One of the important long-range effects of these reports and letters of Schoolcraft was an increased consciousness of the importance of the Keweenaw in the minds of those in Washington. They began to call more vigorously for extinguishing the Indian claims to the region. Thomas Hart Benton, the senator from Missouri from 1820 to 1851, was particularly interested in seeing the mines developed. A champion of westward expansion, on several occasions he pursued the issue of the Indian claims to the mineral lands on Lake Superior.

Perhaps prompted by the Schoolcraft report, he introduced a bill to that effect early in 1823.[60]

One of those from whom Benton sought advice was Governor Cass, who was known particularly for his knowledge of and general sympathy for the Indians. Cass, in responding to Benton's inquiry, was reluctant to approve a complete extinction of the Indian title, believing that "it would be useless to the United States and injurious to the Indians."[61] Because Cass thought it was clear that mining would occur on only a very small part of the land, he believed that when such mining tracts were located they should be bargained for individually. Such a scheme would therefore "leave to the Indians the possession of almost all their country."

The issue kept coming up, and in 1824 Cass once more was called on by Benton to "state the advantages which would attend the purchase by the United States of the country upon Lake Superior, where copper has been found."[62] Again Cass expressed his previous opinion that the elimination of the Indian claims would be a mistake, and that individual parcels could be obtained as necessary. He believed that "little would be gained by an attempt to push permanent settlements into those regions" for the country was so inhospitable. Cass estimated the cost of obtaining the specific lands needed at ten thousand dollars or even less, but he was concerned with the effects any cession might have on the welfare of the Indians. Although he clearly recognized that development of the region was inevitable and even desirable, Cass's sympathies for the Indians remained: "But it is due to the character of our country, and to the feelings of our citizens, that, in our negotiations with these wretched people, we should remember our own strength and wealth, and their weakness and poverty. That we should look back upon what they have lost, and we have gained, and never forget the great moral debt we owe them."[63]

Meanwhile, public interest in the district continued to increase. Benton's bill to obtain title to the lands along the Ontonagon River was published in a Detroit newspaper, and the beneficial effects it might have were not unnoticed. "This enterprise of the National Government will be productive of important consequences to this Territory. . . . The working of these mines will create an active and profitable commerce on Lake Superior, in which this peninsula . . . must largely participate."[64] The ongoing development of the copper district, however, was frequently intertwined with the state of Indian affairs in the region, and in 1826 another such occasion arose.

The important Treaty of Prairie du Chien had been concluded with the representatives of many Indian tribes but, as the distance was great, many Chippewa from the upper lakes did not attend the signing. Early in 1826, therefore, Schoolcraft received a letter informing him that he was being placed in charge of organizing a convention of Indians near the head of Lake Superior. There he was to explain the treaty to them and obtain their agreement to its terms.

He apparently did not look upon the job with enthusiasm: "A treaty at Fond du Lac, 500 miles distant, and the throwing of a commissariat department through the lake, is no light task."[65] Another major purpose of the convention, however, was to obtain the title to the Indian copper lands. Benton was continuing to press the issue. He wrote Schoolcraft that ten thousand dollars had been appropriated to pay for the takeover, and the meeting was "so that the copper-mine business is arranged."[66] As the time approached, Governor Cass, who was to be part of the governmental party, indicated that speed was essential. He had his eye on bigger things for himself. Perhaps frustrated because he had not made it to the copper boulder in 1820, he had long coveted its removal to Washington, where he was soon to go as secretary of war. The boulder would make an admirable trophy with which to arrive.

In any event, no time was to be lost. Once title to the land on the Ontonagon River was obtained, Cass wrote Schoolcraft, "we must remove the copper-rock, and, therefore, you will have to provide such ropes and blocks as may be necessary." On the fourth of July, Cass and Thomas L. McKenney, the commissioner of Indian affairs, both having been appointed by President John Quincy Adams, arrived at the Sault to join Schoolcraft. After six days of final preparations all was in readiness. "Jason," wrote Schoolcraft, "could not have been more busy in preparing for his famous expedition to Argos." They set out in fine weather, with the military contingent alone consisting of sixty-two men. To the accompaniment of flags and music, Lake Superior "yielded before Anglo-Saxon power."[67]

Eighteen days of travel brought them to Fond du Lac. On the way, they camped at the mouth of the Ontonagon, where, some twenty miles upriver, the copper mass lay. McKenney, like so many before him, had fallen under its spell. "Its weight is estimated to be nearly *three thousand pounds,*" he enthusiastically noted in his journal. "We hope to possess ourselves of this great curiosity; and if we succeed, I shall take it home with me." Or, perhaps as he returned to reality, "I will be able to carry with me some specimens."[68] First, however, there was the matter of the treaty.

At Fond du Lac the Indians were informed of what was desired: "We also wish that you would allow your great father to look through the country, and take such copper as he may find. This copper does you no good, and it would be useful to us to make into kettles, buttons, bells, and a great many other things."[69] The Indians had their chance to speak in response. Chief Plover from the Ontonagon implied that he wasn't sure what they were after: "Fathers,—I have no knowledge of any copper in my country. There is a rock there. I met some of your people in search of it. I told them if they took it to steal it, and not to let me catch them." Another unnamed chief from the Ontonagon, however, perhaps saw their dilemma more clearly: "This, Fathers, is the property of no one man. It belongs alike to us all. It was put there by the Great Spirit, and it is ours. . . . If you take this rock, Fathers, the benefit to be derived from its sale, must be extended to our children, who are now but this high. For ourselves, we care but little. We are old and nearly worn out. But our children must be provided for."[70] Finally, amid much pomp and circumstance, the treaty was concluded.

The Indians got some land parcels, cash, a school, and special medals to wear. The medals, distributed by Schoolcraft, "have on one side of them your great father's face, and on the other side is his pipe, his peace hatchet, and his hand. It is a new heart, an American heart."[71] Of the nine articles in the treaty, Article III is significant here:

> The Chippeway tribe grant to the government of the United States the right to search for, and carry away, any metals or minerals from any part of their country. But this grant is not to affect the title of the land, nor the existing jurisdiction over it.[72]

This agreement followed the earlier recommendations of Governor Cass, in that the title of the land was not ceded, but only the mineral rights. But then, the issue had never really been in doubt anyway. Five days *before* the treaty was signed, two barges loaded with equipment and twenty men had departed, bound for the Ontonagon River.[73]

George F. Porter and a Colonel Clemens were assigned the task of removing the boulder. They were not the first to try, but had come prepared with wheels, ropes, and a score of helpers. Proceeding up the river, they left one of their boats, along with most of the provisions, at the first rapids. They spent another day and a half dragging their second boat, with great difficulty, up the higher rapids. After going an estimated thirty miles, they abandoned the boat and struck out across very difficult country by foot. Finally they reached their goal, "this remarkable specimen of virgin copper."

It was between steep valley walls and above three rapids that dropped seventy feet. They soon realized that even with their equipment there was not the slightest chance that they could ever move it to the lakeshore. Wanting to bring back at least something to show for their efforts, they built a huge bonfire around the boulder with all the dry wood they could find. Dashing water on it, they hoped to crack off some samples, an "attempt to mutilate and falsify the noblest specimen of native copper on the globe," Schoolcraft wrote later.[74] Porter simply noted that with their means and time it was impossible for them to move it either by land or by the river, and they were obliged to return to the lake empty-handed.[75] There they met the rest of the party journeying back to the Sault from Fond du Lac. McKenney was mortified to learn of their failure, and he gazed longingly up the river as they passed, his one chance now gone. Treaty or no treaty, although beaten, bent, broken, and now burnt, the copper boulder was still the master of its fate.

The treaty of 1826 did not, despite all the vigor with which it had been pursued, result in any immediate rush to begin mining. The treaty granted the mineral rights "to the government of the United States," but not to private individuals. The government was not about to get itself into the mining business directly, and the land itself remained in Indian hands. The government, therefore, held the key to any future mining efforts, but as long as title to the land was not held, and given the absence of compelling economic pressures, active attempts at mining remained unlikely.

The state of geological knowledge about 1830 also contributed to the uncertainties. I have emphasized the growth of governmental and public awareness of the district, and the significant contributions of Schoolcraft to a more accurate appraisal of its potential. Yet with a few minor exceptions virtually all the copper masses discovered had not been found in the solid rock of the earth. They were lying on or near the surface as part of loose, unconsolidated material. Further, there was general agreement that such masses had been transported by some geological agency to where they had been found, perhaps even from great distances away.

The Ontonagon boulder was the classic example. All investigations had failed to find any more copper in the nearby bedrock. Where the boulder had come from originally, or where such native copper or perhaps other significant copper ores might be found *within the rock itself* was an unanswered question, and without that answer exploitation of the copper of the Keweenaw had to be an unfulfilled and uncertain potentiality.

However, another important step was about to be taken. Back in his position as Indian agent Schoolcraft was getting restless. He had long been enamored of the possibility of discovering the true source of the Mississippi River. Steps toward that discovery had already been taken on the Cass expedition of 1820, but that party had not reached the point of origin of the great river. Even though it had not been one of the initial stated objectives of that expedition, the search for that source had become, in Schoolcraft's mind at least, the primary goal of the earlier trip. Ever since, in an age that greatly honored the explorers of unknown lands and the heroes who were pushing at the frontiers of geographical knowledge, he had wished to achieve a measure of immortality by finding the true source of the Mississippi. It was unlikely that an expedition for that sole purpose would be financed, but a new opportunity soon presented itself.

By 1830 there had been for many years almost constant trouble between the Sioux and Chippewa Indians in the region west of the Great Lakes. The earlier treaties had not quelled the hostilities, and tensions were again running high. In August of that year the war department directed Governor Cass to ask Schoolcraft, the Indian agent for the upper lakes, to try to settle this long-lasting dispute. Schoolcraft soon realized that this was the golden opportunity for which he had been waiting.[76] He made the most of the chance by organizing expeditions to the north and west of the lakes in both 1831 and 1832. On the second of these he finally attained his objective, although it was accomplished under the guise of visiting and pacifying the Indian tribes.

It was too late in 1830 to begin, so Schoolcraft once again spent the winter planning an expedition. He sent a letter to the Indian chiefs whose actions were disturbing the peace, reminding them of their obligations under the earlier treaties and warning them that he was coming: "My Children, Your great Father at Washington has heard of the war."[77] Although the intent of the mission was entirely Indian related, provision was made to employ a person who would make notes on the nature of the soil and the geology, mineralogy, and natural history of the region to be visited. Schoolcraft also needed a physician for the journey, to serve both the members of the expedition and the Indians they would be meeting along the way. Back in 1820 he had been chosen to accompany the Cass expedition as its mineralogist. Now Schoolcraft was the leader, and the choice in turn was his to make. It was to be a momentous decision for the future of the geology of the Lake Superior region generally and the Keweenaw copper district in particular. The man he selected to accompany the expedition, only twenty-one years of age and a very recent arrival in the Michigan Territory, was Douglass Houghton.

4

DOUGLASS HOUGHTON—
COPPER FINDS ITS COLUMBUS

Down through the years, Douglass Houghton and Keweenaw copper have come to be almost synonymous. No person before or since has ever been so closely associated in the public mind with this unique district. He was invited to come to Michigan when it was still a territory, part of the great, sparsely settled west. At his tragic death only fifteen years later, he was one of the most prominent and best-loved men in the state and enjoyed a national reputation. Active in many areas of public life as well as being Michigan's first state geologist, his most lasting fame came as a result of his fundamental work in the Keweenaw copper region. He has been called both the George Washington and the Columbus of the district, as well as "the father of copper mining in the United States."[1]

Early Life and Education

He was born September 21, 1809, in Troy, New York, the son of Jacob Houghton and Mary Lydia Douglass, and he bore both family names (fig. 12). The family soon moved to Fredonia in western New York state where Jacob, a lawyer, prospered and eventually became judge of the county. He was enthusiastic about the education of his children, and Douglass, one of five brothers and two sisters, grew up in a closely knit, cultured home. Its country setting provided an ideal opportunity

Fig. 12. Douglass Houghton. One of several portraits of Houghton by Alvah Bradish, but probably the only one made during the geologist's lifetime, now at the Michigan Geological Survey in Lansing. Courtesy Michigan Historical Collections, Bentley Historical Library, The University of Michigan.

for early contacts with the natural world, and he was soon drawn to botany and geology.

Although of diminutive stature, he exhibited early the nervous energy, the emphasis on action as opposed to contemplation, that was to characterize his entire life. He was inquisitive and curious, and a youthful experiment with gunpowder left him with permanent facial scars. Although he was a voracious reader, his mental bent was clearly toward the practical and scientific, as opposed to the classical and refined, a characteristic noted by many in his later life. These traits undoubtedly contributed greatly to his success in a frontier environment. A recent brief account of his career notes that his "combination of pragmatic and scientific qualities was peculiarly suited to early nineteenth century America."[2]

The Houghton children were among the first students to enroll in the new Fredonia Academy, and their efforts were strongly encouraged at home. Still in his middle teens, while at the academy at Fredonia Douglass also studied medicine independently under a doctor friend of the family, thereby laying the basis for the profession that would endear him to so many in Michigan as "the little Doctor." His studiousness and intelligence evidently attracted the attention of others, although, perhaps characteristically, he did not like Latin despite its significance to a scientific and medical course of study. Early in 1829 he was selected to continue his education at the Rensselaer School at Troy, New York.

In 1824 Stephen Van Rensselaer had established a school at Troy, New York, as an outgrowth of his long-standing interest in science generally and geology in particular. This institution was to have a great influence on the development of geology in America during the remainder of the nineteenth century.[3] Amos Eaton, a prominent early American geologist for whom the 1820s have been called "the Eatonian Era," was appointed to be the senior professor at the school.[4] There, Eaton introduced a course of study that strongly emphasized science, in contrast to the education available at many other institutions in the country, thus taking a significant step toward a pedagogical revolution in higher education.

The educational philosophy of the institution, which was effectively that of Eaton himself, was clearly articulated. "The most distinctive character . . . consists in giving the pupil the place of the teacher. . . . Taking advantage of this principle, students of Rensselaer *learn* by giving experimental and demonstrative lectures." Practical exercises abounded, always with an eye to "direct application to a useful purpose," and emphasis was placed on work in the field.[5]

Many prominent geologists of the remainder of the century had ties with the Rensselaer School, either as students or faculty. Among them were Lewis Beck, George Cook, Ebenezer Emmons, and James Hall, one of the most influential geologists in American history. By 1860 seven state geological surveys were or had been headed by Rensselaer graduates, and many other graduates were active in the surveys at lower levels. Although Houghton's scientific education was broadly based and not narrowly focused on geology, his education at Troy and other associations he made clearly were taking him into the newer "professional" tradition taking shape in American geology. The day of the largely self-taught was passing.

The education available at Troy was ideally suited to Houghton's personality and natural inclinations. He "presented his credentials" on April 14, 1829, and found that the school "more than equals my expectations." In a letter to his brother he left an enlightening description of the laboratories and educational procedures as he described the rigorous study day that began at half past four in the morning.[6] By October 1829, after only about six months at Troy, he received his bachelor of arts degree. Clearly he had learned his lessons well, and he was asked to remain at the school as an assistant instructor. The following February he was appointed by Eaton to be an assistant professor in chemistry and natural history.[7]

For the summer of 1830 Eaton planned a significant teaching innovation for the students at Rensselaer: a full fledged-field program in the form of a ten-week "Summer Term of Travelling Instruction."[8] The students "will be conveyed by a flotilla of towed canal boats." The largest boat was equipped with a reading room, along with

> suitable chemical and philosophical apparatus, and cabinets in mineralogy and geology. Applications will be made, by direct inspection of rocks and minerals in place, plants and minute animals in their native localities, the works of the engineer in actual operation, the labors of the agriculturist, &c. Students of the course will be taught the method of procuring specimens in natural history, and required to make collections of whatever is interesting upon the route.

Houghton went along as second in command. Upon Eaton's illness he finished out the term with a three-week session on chemistry back at Troy, after finding that the chemical experiments could not be conducted well on the boat.[9] Here was indeed a new and farsighted view of education and the part that science, and particularly field experiences and practical applications, should be in it.

The journal of the trip was published later in the *American Journal of Science:* "Wednesday, 23—Collected specimens of Basalt in most of its varieties; Thursday, 24—Visit the extensive vein of copper pyrites; Friday, 25—Take the steam boat for Catskill; July 1st. Thursday—Embark on board the canal boat; Friday, 2,—The students visit the Rev. Dr. Nott at Union College, he being president." At Salina, "the manufacture of salt by boiling and solar evaporation is very interesting."[10] And so on. It sounds fascinating, although Eaton was a harsh taskmaster. Later, Houghton's confidence in him was severely shaken when student resistance erupted in "complete rebellion."[11]

Both Eaton and Ebenezer Emmons, the junior professor at Rensselaer, wrote textbooks that were used at the school. Emmons's *Manual of Mineralogy and Geology* of 1826, which was second only to Cleaveland's book of 1816 in American mineralogy, and Eaton's *Geological Textbook* of 1830, written while Houghton was his assistant, are probably representative of the geological education the students at Rensselaer were receiving. Eaton's book opens with his concept of geology: "The science of Geology consists in a systematic arrangement of facts, explaining the structure of the earth."[12] These words, fitting in so well with his inclinations, could perhaps be taken as a brief statement of Houghton's philosophy of geology.

More intriguing are the curious words with which Eaton began the preface to his work, the very first words encountered on opening the book: "Every geologist is, probably, more or less misled by theory." It is very likely that these sentiments were a part of the approach to nature that Eaton tried to instill in his students. Houghton clearly learned the lesson well, and those words do capture with amazing fidelity the essence of what Houghton later *claimed* was his approach to science, and to geology in particular. At first glance this philosophy would seem to associate Houghton more closely with the earlier amateur tradition, rather than with the newly emerging professional tradition. Yet it is evident that theoretical considerations were playing an ever increasing role in the geological thought of the day. Furthermore, as the events examined below will make clear, there was a distinct difference between what Houghton honestly believed his scientific philosophy was, and the way he actually went about his scientific work in the copper region, a difference that has not always been appreciated in other accounts of his efforts.

Also significant, in the light of Houghton's future experiences with the copper in Michigan, are Eaton's views about mineral districts generally and the appropriate techniques that would be called for in moving into a new or unexplored region: "Geology teaches us that minerals which are associated in one district of the country are associated in the same order in all other districts. Hence the experience of the miner and the quarry-man in any country may be applied in searching for useful minerals in all countries."[13] Eaton was a particular proponent of this doctrine or principle of analogy. In many ways it seems to be a most obvious approach to nature. What could be more reasonable than to use that which is known in trying to understand that which is not known? Once again Houghton learned well, and it is evident that this manner of thinking played a specific and important role as Houghton tried to unravel and understand the nature of the copper district. The problem was, of course, that the Keweenaw district turned out to be, as they could not have known at the time, distinctly nonanalogous.

The circumstances that brought Houghton west to Michigan have become part of the folklore surrounding him. The long Michigan winters had been taking their toll on the local citizenry at Detroit, and with the arrival of the autumn of 1830 the prospect of facing another dreary cold winter indoors with little to do was not encouraging. With the support of a number of the leading citizens of the territory, a Detroit newspaper began a campaign to bring a lecturer on scientific subjects to the city, hoping to "prevent our frisky old bachelors and gay young belles, puny young beaux and prim old maids from suffering so much ennui as to cut their throats during the long season."[14] They took their inquiry to Amos Eaton, "the leading scientific educator in this country," for suggestions, and he unhesitatingly recommended the youthful Houghton, despite the reservations expressed by the Michigan inquirers because of his youth. Houghton was enthusiastic about the assignment, and he quickly prepared to make the journey. He arrived in Detroit in early November 1830, just turned twenty-one years old and with "only a dime in his pocket." He nevertheless immediately proceeded to immerse himself completely in the life of the territory.[15]

He was about five feet three inches tall, somewhat delicate, with a quick, nervous demeanor, a great storyteller, always courteous, at home with all social classes. It is difficult to find even a hint of any disaffection with him among his contemporaries. His lectures, despite a degree of hesitation in his speech, were given on a paid-subscription basis to turn-away crowds. Prominent citizens soon began to gather around him "like bees around honey," and his friends among the younger men of the city soon began to style themselves "the Houghton Boys."[16] The next spring he returned briefly to New York to be licensed as a physician, preparing to begin his career as "the little Doctor."

The hand of fate intervened, however, in the form of Henry Schoolcraft, who at that very moment was looking for a physician and naturalist to accompany him on the expedition he was planning for the coming summer. Evidently impressed by Houghton's qualifications and personality, he offered the position to the young doctor. So, after only a few months in the Michigan Territory, Houghton found himself bound for the north country and his first look at the region that became the focus of so many of his future efforts and, too soon, the scene of his death.

The Expedition of 1831

Five days were spent by the members of the expedition traveling from Detroit to the Sault, where the final preparations for the push westward were made. From there they set out about the twentieth of June, 1831, with Schoolcraft, Houghton, and Melancthon Woolsey, the secretary of the expedition, occupying one large canoe manned by eight voyageurs for extra speed. This allowed them the freedom to make investigations without detaining the remainder of the flotilla. Ahead of them was a journey of some two thousand miles, which they completed in seventy-two days, coasting the south shore of Lake Superior, visiting Indian tribes in Michigan, and the regions of Wisconsin and northern Illinois, and finally returning up Lake Michigan back to the Sault.[17] Much later, Schoolcraft was to recall this journey in idyllic terms, speaking of the three men as the harmonious strings of a harp: "The sainted and scene-loving Woolsey—the self-poised and amiable Houghton, just broke loose from the initial struggles of life to luxuriate on the geological smiles of the face of nature in this scene—ah! where are they? Death has laid his cold hand on them."[18]

It seems likely that Houghton would have heard the stories of Lake Superior copper that had circulated so widely after Schoolcraft's earlier trip, but there is no reason to suspect that he had any interest in the copper before he came to the Michigan Territory. However, Governor Cass, perhaps recalling his own earlier, unsuccessful efforts to reach the copper boulder and to have it moved, seems to have asked Houghton to visit it and report back to him, even though the main purpose of the expedition concerned Indian affairs.

Houghton enthusiastically entered into the activities of the expedition, making geological plans even before they left the Sault. In a letter written to his brother from there, he said that he was going to make a quick trip to Point Iroquois on the lake "for the purpose of laying the foundation of a Geological map which I am to project."[19] It would seem reasonable to assume that Schoolcraft would have freely shared what he knew of the geology and mineralogy of the region while they passed the many miles together in the same canoe along the southern shore of the lake.

Six days were used in coasting the long route around the tip of Keweenaw Point. The group paused frequently to examine rocks and blast for specimens, and occasionally journey inland for some distance. They examined the "roche verte" near the northern tip of the peninsula, that vein of green carbonate and silicic ores so often noted by the voyageurs but which Schoolcraft did not see on the Cass expedition of 1820. They found that this vein, besides the green copper-bearing minerals, contained some black oxide of copper where it continued inland. They also found a vein that contained native copper as well as some other minerals of copper compounds. Continuing along the coast, they soon reached the Ontonagon River and its copper boulder.

On their arrival, Schoolcraft elected to stay at the mouth of the river and wait; he had seen enough on his previous travels. The priorities of his life had shifted, and mineralogical and geological concerns were not as important to him as they once were.[20] His decision, seen in retrospect, was symbolic. The torch was being passed, and he would now yield to youth. Houghton, along with a party of thirteen, eagerly went up the river where he got his first look at the copper wonder of the world. He returned the next day with some large samples and a mind brimming with ideas.[21] They continued on west and south to visit the Indian tribes and to get on with their primary responsibilities, but as they went, Houghton collected plants extensively. At this time botany evidently held an attraction for him nearly equal to that of geology. His efforts in this direction, continued the next year, constituted the first systematic analysis of the plants of the region.

Houghton returned to Detroit on the eighteenth of October. His arrival, and the significance of his observations on the copper, attracted the attention of the local newspapers almost before he had stepped off the boat. News from the north, especially if it involved the possibility of turning a profit, evidently sold papers. The very next day it was reported that "the discovery of a new and extensive copper mine at Point Kewewena [sic] is important to the commercial interests of this Territory, and it will undoubtedly become a source of wealth to the country."[22] Neither were the implications lost on the businessmen of the city. On the twenty-eighth of October, "a large and respectable meeting of the citizens of Detroit" was held to write a petition to Congress on internal improvements of the territory. Among the improvements called for was a canal around the rapids at the Sault to facilitate shipping, and also requested was

> the opening and working of the extensive copper mines on, or near, the river Ontonagon. The United States expends annually a large sum of money for the purchase of foreign copper, for the use of the navy, and for the merchant service. The amount thus paid for imported copper, might not only be saved, but a sufficient quantity of that valuable article procured for exportation.[23]

The rapids at the Sault had long been a bottleneck to access to the upper lake, as supplies and passengers had to be off-loaded below the rapids and portaged to another vessel over twenty feet higher on Lake Superior. This call for a canal became a continuing theme in future years until the project was finally completed in 1855.

On November 14, 1831, after returning to his home at Fredonia, New York, Houghton filed his report. It was in the form of a letter titled "A Report on the Existence of Copper in the Geological Basin of Lake Superior," and was written to Lewis Cass, who in the meantime had gone to Washington as the new secretary of war in President Jackson's cabinet.[24]

Houghton began, in his characteristically conservative manner, by acknowledging the excessive attention the copper had already received: "It is without doubt true that this subject has long been viewed with an interest far beyond its actual merit," with every loose mass being considered the location of a possible mine. On the other hand, "it is no less certain that a greater quantity of insulated [isolated] native copper has been discovered upon the borders of Lake Superior, than in any other equal portion of North America." The traprock on the east side of Keweenaw Point was rich in a variety of minerals, including "several of the ores of copper." This trap also appeared on the west side, with copper black ore disseminated through the rock, along with native copper. This led him to conclude that the trap formation "extends from one side of Keweena [sic] Point to the other, and that a range of thickly wooded hills, which traverses the point, is based upon, if not formed of, that rock."

This range of rugged hills that runs the length of the peninsula is its dominating feature, but what rock it was made up of had long been an unanswered question. Earlier, Schoolcraft had speculated that it was probably composed of granite, which made up the district's "nucleus" or "spine."[25] This may well have been an echo of the traditional view that granite, according to the old Wernerian geological scheme, was the rock to be expected at the center of any major uplifted region. Houghton here recognized that it consisted instead of traprock in a variety of forms: "compact granular, amygdaloidal, and toadstone [slaglike]." He also found that, as they rounded the point, the trap gave way to a "crag" [conglomerate], which overlay the trap, and it was this crag that contained the famous "green rock" vein. The trap then appeared again a few miles farther west. In this analysis, important steps toward understanding the basic geological structure of the district were taken.

As for the Ontonagon boulder, Houghton said that it was a completely isolated mass that had a small quantity of rock still adhering that "is evidently a dark colored serpentine." The only rock he found in place near the boulder was the common red sandstone of the south shore, which did not seem to contain copper anywhere. Therefore, the still visible traces of the attempt by Alexander Henry to put down a mine at that location were, he bluntly declared, "a memento of ignorance and folly." It is clear, he said further, why the earlier mining attempts were failures, for "no attempts were made to learn the original source of the metal which was discovered, and thus, while the attention was drawn to insulated [isolated] masses, the ores, ordinary in appearance, but more important *in situ*, were neglected."[26] There was little evidence at the site of the boulder "that would enable us to judge of its original geological position." Then, however, Houghton applied some simple but astute geological reasoning.

A few days earlier, back up the coast toward Keweenaw Point, he had seen dark traprock containing some native copper outcropping along the shoreline, but this traprock had disappeared from the shoreline, receding inland as they proceeded southwest. Now, apparently on the return journey to the rivermouth, he went up the eastern fork of the river for some distance, where "I discovered small waterworn masses of trap-rock, [with] occasionally associated minute specks of serpentine, in some respects resembling that which is attached to the large mass of copper." We may therefore conclude, he wrote, "that the trap formation which appears on Lake Superior east of the Ontonagon River, crosses this section of country at or near the source of that river." This indicated that the traprock ran along the entire length of the Keweenaw peninsula, and also that it was the original source from which the copper boulder had come.[27]

The origin of the copper masses scattered over the country had long been a mystery. Schoolcraft had made some first steps toward an answer, but for Houghton the mystery was solved:

> After having duly considered the facts which are presented, I would not hesitate to offer, as an opinion, that the trap-rock formation was the original source of the masses of copper which have been observed in the country bordering on Lake Superior; and that at the present day, examination for the ores of copper could not be made in that country with hopes of success, except in the trap-rock itself; which rock is not certainly known to exist upon any place upon Lake Superior, other than Keweena [sic] Point. . . . What quantity of ore the trap-rock of Keweena Point may be capable of producing, can only be determined by minute and laborious examination. The indications which were presented by a hasty investigation are here embodied.

This report, seen in its totality and in the light of the circumstances behind its writing, is a remarkable document, one that represented a major step forward in understanding the nature and structure of the district. It achieved great exposure, and was published as a congressional document the next year,[28] and again in Schoolcraft's popular account of the discovery of the source of the Mississippi that appeared in 1834.[29] Thus there began, in the public mind, the close association of Douglass Houghton and Keweenaw copper that was to continue for the remainder of his life.

After their return to Detroit, he went back to his home state of New York, apparently having no further ties to Michigan. He seems to have inquired of Secretary Cass about the possibility of a position with the medical staff of the army, and he was soon anticipating an appointment to an army post.

However, his first introduction to exploring on a large scale had been much to his liking. While awaiting a response from the war department, he was invited by Lucius Lyon, a government surveyor and prominent citizen of the Michigan Territory, to be the naturalist on an expedition that was to spend the summer of 1832 trying to settle Indian boundary disputes west of the Mississippi River. Houghton eagerly accepted the position.

The route was to be along the Iowa, Des Moines, and Missouri rivers, which would, Houghton believed, "undoubtedly afford a rich field for investigation, particularly in the field of botany." It further offered, not incidentally, the possibility of "a handsome salary."[30] John Torrey of New York, the country's leading botanist of the time, was much interested in the plants collected on the 1831 trip. Houghton informed him that the newly planned trip would be of "far greater interest in a botanical point of view," and that he would be glad to try to get any plants Torrey might want from this "constant botanical feast."[31] Just before he was to leave, however, fate once again intervened in the familiar form of Henry Schoolcraft.

The Expedition of 1832

Schoolcraft was most anxious to reach the true source of the Mississippi River, but the expedition of the previous summer was too limited in scope to permit attaining that goal. Accordingly, he managed to convince the appropriate governmental authorities, with the tacit approval of Secretary Cass, that another visit to the Indians was needed to help settle their differences, but this time especially to those tribes in the region of the upper Mississippi valley. With the approval of this new expedition a surgeon and naturalist was once again needed, and in view of his outstanding work the previous summer Houghton was the logical choice. Despite his previous commitment to Lyon, he jumped at the chance. "I do not consider it will be a breach of faith to join you as you propose," he wrote to Schoolcraft. His friends had been trying to get him to settle down in Fredonia to practice medicine, but he termed that possibility a "dog's life; they can but little realize the pleasure of a mental feast upon the hidden treasures of nature."[32]

And once again Schoolcraft faced planning a major expedition. Lt. James Allen, named director of the military escort, was to keep a journal that would include observations on geology and mineralogy. Houghton, as surgeon, was to meet an important objective of the trip by vaccinating the Indians against the smallpox that had been ravaging the tribes. He was delayed by a lawsuit in which he was the plaintiff but made it to the Sault in time for the departure. The group of thirty-five men, along with all the necessary supplies, left on June 7, 1832, for what proved to be a nearly three thousand-mile journey.[33] "If I do not see the 'veritable source' of the Mississippi, this time," wrote Schoolcraft, "it will not be from a want of the intention." He needn't have worried.

Houghton's primary official responsibility was the vaccination of the Indians, a task he carried out faithfully but which became tiresome at times. At Fond Du Lac he vaccinated 240 in one evening. He also used his surgical skills when needed: "Tomorrow morning I expect to take up the brachial artery in a case of advanced aneurism," he wrote his brother.[34] His plant collecting also took much of his time and interest, and he was gratified by a letter from Torrey confirming some of his identifications of the previous season. Torrey was flattered that Houghton was planning to spend the next winter in New York to get further experience in medical topics, and said in a letter to Schoolcraft that he would do anything to aid Houghton in his botanical studies, "but I would on no account deprive him of the honor he has acquired with so much labour & at so much risk."[35] Houghton's scientific reputation was on the rise.

As they passed westward along the south shore, Allen, the head of the military escort, made many geological and mineralogical notes, which achieved considerable circulation after his return and which indicate that he too was an astute observer.[36] As they approached the Keweenaw, Schoolcraft and Houghton took the portage route, having rounded the point the previous year, and Allen accompanied the larger boats on the long route around. He noted the famed "green rock" near "Point Keewaywenon" and gave a concise description of several copper veins in the trap along the shore to the west that had been noticed the previous year, commenting that they looked promising.

He rejoined Schoolcraft and Houghton at the head of the portage, and Allen and Houghton then left the main party and sailed on ahead, hoping to reach the Ontonagon River early to make a quick trip to the boulder. The wind died, however, and they were soon overtaken. Because it would have delayed the group for perhaps two days, they "were induced to abandon the project." In any case, Allen noted that it "has been so often visited and described, that it has lost a great part of the interest and curiosity which it at first excited; and the many unsuccessful searches for copper mines, in its vicinity, have nearly exploded the theory of their existence."[37] Houghton and Allen did agree, however, that they would meet and examine some of the features of the south shore of the lake in a more deliberate manner on their return trip.

In August, after the source of the Mississippi had indeed been located and the official business of the expedition was concluded, and after Allen had had a most difficult journey that resulted in his pressing the serious charge of abandonment against Schoolcraft, the party broke up. Schoolcraft went on ahead and Houghton waited for Allen, as they had previously arranged, so that they could journey together back to the Sault. At the Ontonagon River they had hoped to find Indian guides to take them to the boulder, but the village was abandoned for the summer. They did find a small canoe hidden in the bushes, so Allen, Houghton, and two men headed upriver, camping overnight on the way and nearly being eaten alive by sand flies and mosquitoes.

As usual, at the forks of the river they struck off overland some miles westward and found, significantly, some traprock in place in the side of a small hill. As they had passed the Porcupine Mountains on the initial, westward leg of the journey, Allen had casually remarked that they were probably granite, in accord with the conventional wisdom of the day. This observation of traprock, therefore, contradicted that supposition, a fact Allen specifically mentioned in his journal. Another day of wandering, trying to find the trail, another night on the ground, and finally, after an "annoying and difficult journey," they reached their goal.

Houghton thought that the boulder looked bigger than he had remembered it from the previous year, and Allen estimated its weight at four to five tons, concluding that it was 'the *largest mass of native copper ever found.*" With difficulty they hacked off some samples and then carefully examined the immediate vicinity for any other native copper or copper ores, this, wrote Allen, being one of the objects of the visit. They found "not a particle or trace of either."[38]

Back at the lake, they resumed their journey around the peninsula, stopping first at the copper veins found the previous year and mentioned by Allen on the earlier passage west. There they took samples of native copper, copper green, and copper black, found as small specks in the rock. A few miles farther brought them back to the "green rock" where they blasted and uncovered a vein of completely pure black copper oxide, one of the richest ores of copper. The vein was only about six inches wide, but it increased as it descended into the rock. They speculated that the vein probably descended all the way into the underlying trap, which was the bedrock farther inland, and concluded that the locality certainly merited further investigation. Proceeding eastward along the coast through occasionally heavy weather, they reached the Sault and the termination of the expedition on the twenty-fifth of August.

With the safe return of the members of the traveling party to civilization Schoolcraft was widely hailed as the discoverer of the true source of the Mississippi. Through his published accounts of the expedition and the journals kept by the principal figures, the results of the journey became very widely known. Allen's journal, which included the geological observations he made with Houghton, was entered as a congressional document.[39] Houghton had written several letters back to a Detroit newspaper, which thereby kept its readers informed of the expedition's progress.[40] The paper later editorialized that the observations made on the expedition indicated that copper ore would be found in abundance and would "reward future exertions. It may have a mark'd influence on the coming prosperity of Lake Superior."[41] Houghton also wrote to Torrey about the plants he had collected, and apologized for not being able to visit New York as planned. He was also thinking of some way of organizing the geological observations he had made: "I am most anxious to complete, for publication, a Geological Map of Michigan, N. West, & a part of Missouri Terrs. [sic]."[42]

The Quiet Years

And yet, after such a promising beginning, over the next several years Houghton's interests moved in different directions. His temperament was that of a man of action, eager to move on to new things. He returned to Detroit and finally decided to take up his medical profession, which soon became profitable both financially and socially. In 1833 he married his childhood friend from Fredonia, Harriet Stevens, and settled down to become a solid and active member of the high society of the Michigan Territory: "our Dr. Houghton, the little doctor." He also had a good eye for the ways money could be made in a frontier society, and he became deeply involved in real estate investment and land speculation, buying land from the government and quickly reselling at a large profit. "I purchased yesterday," he wrote his father, "2,600 acres of pine lands which cost $1.25 per acre, worth now from $5.00 to $10.00 per acre."[43]

As the years passed, so much of his time was being taken up by financial investments that he, the most successful doctor in the territory, found it necessary to stop his regular practice of medicine, "as I have stretched my arms so wide that my profession was a loss rather than a profit." He still lectured locally on geological topics, and he was forced to seek larger quarters as his house became a virtual museum of natural history, but until the Michigan statehood year of 1837 his life was that of a successful and prosperous private citizen.

In the meantime, little was happening in attempts to develop the copper district. The transportation bottleneck at the rapids of Sault Ste. Marie made it difficult to move supplies and equipment into Lake Superior, and the winters were severe. The treaty of 1826 had allowed prospecting for minerals, but ownership of the land itself remained with the Indians, whose sporadic warfare made for unsettled conditions. True, the district was administratively a part of the Michigan Territory, but it was a very distant part. Statehood for Michigan was approaching, but to most residents, who were concentrated in the area around Detroit, Michigan meant the lower part of the Lower Peninsula and, perhaps reluctantly, the region around the Sault, but beyond that was the wild unknown.

Strangely, after all the attention from Washington for so many years and all the favorable publicity for so long, the Keweenaw copper district became embroiled, indirectly, in the bitter controversy that enveloped the Michigan statehood period. For a while it seemed destined to become part of the stepchild that no one wanted. Had things gone just slightly differently, in the direction that virtually all residents of the Michigan Territory had wanted them to go, the copper district would have ended up as part of the state of Wisconsin. In that case, Houghton's task as the first state geologist of Michigan would have been made much easier, and the subsequent history of the district might have been quite different.

The territorial boundaries drawn and redrawn in the region northwest of the Ohio River were basically lines of expediency. They were attempts to bring the area under some sort of administrative control until the time that the "three but no more than five" states prescribed by the Northwest Ordinance were created within the region. With the admission of Ohio to the union in 1803 the future conflict that came to be known as the Toledo War was assured. The Ordinance of 1787 provided clearly that the boundary between the northern and southern states in the region would be a line drawn east and west through the southern extremity of Lake Michigan. This would have, without question, put the valuable harbor of Toledo, which was at the mouth of the Miami [Maumee] River on Lake Erie, in the state of Michigan. At that time, however, the latitude of the southern tip of Lake Michigan was poorly known. Covering itself, Ohio declared in its constitution that the official boundary could in no case touch Lake Erie south of the northernmost cape of the Miami Bay. Controversy over the disputed strip of land that lay between Michigan and Ohio, the "Toledo Strip," dominated the entire Michigan statehood period, and the fate of the copper district unknowingly hung in the balances.[44]

When first created in 1805, the Michigan Territory had as its western boundary a line passing northwardly through the middle of Lake Michigan to the lake's most northerly point, and from there due north to the international boundary in the middle of Lake Superior. This meant that the eastern end of what is now the Upper Peninsula, including Sault Ste. Marie and the Straits of Mackinac, was within the territorial boundaries, but the copper district was not. These boundaries changed frequently, however, so that as late as 1834 the Michigan Territory extended so far west that it included portions of what is now Iowa and the Dakotas. As Michigan approached statehood, one thing was clear in the minds of almost all of its citizens. The part of its territory that was to become a state did *not* include the western part of what is now the Upper Peninsula, but it definitely *did* include the Toledo strip. No matter. Ohio, already a state for thirty years, clearly had the political clout, and Michigan was told it was welcome into the Union, but without the Toledo Strip. Further, Michigan would either have to explicitly agree to these conditions beforehand or else forget about being admitted at all.

The great interest the copper had created in Washington since 1800 has already been noted, and those who had explored that country and knew it best, Cass, Schoolcraft, and Houghton, all had close ties with the Michigan Territory. Also noted was the interest and excitement that news of the district generated within the territory, and the recognition that the copper would certainly play a role in the economic future of the region. Even so, when the time finally came for statehood, the possibility that Michigan might acquire the remainder of the Upper Peninsula, and therefore the copper district as well, generated almost universal, bitter opposition.

For some time there had been agitation among the few residents of the country north of the Straits of Mackinac to be separated from the Michigan Territory and to be attached to the territory to the west. Detroit, it was felt, was too far for effective government.[45] In fact, in 1829 the legislative council of the territory considered asking Congress to attach *all* the territory north of the straits to the Territory of Wisconsin. Schoolcraft was one of those voting in favor of the action, which, had it been taken, would have resulted in the copper district later becoming a part of the state of Wisconsin.[46] The proposal was not submitted, but the agitation for separation by those above the straits continued, while in the Lower Peninsula all eyes were on the strip that was being disputed with Ohio. Michigan was ready to proceed into the Union, wanted or unwanted, and in 1835 it declared itself a state. Lucius Lyon and John Norvell were elected senators, and they proceeded to Washington to take up their (un)official duties. But the boundary question still remained.

William Preston, a senator from South Carolina, may have been the first to suggest that Michigan be given the land now composing the western part of the Upper Peninsula as compensation for the loss of Toledo. Apparently he felt that attaching it to the Wisconsin Territory would result in too large a state to the west. Schoolcraft, invited to a committee meeting of the Senate, then expressed the opinion that the proposed addition would be far more valuable than the Ohio strip.[47] Lyon evidently was at first opposed, saying that the people of Michigan did not want their state extended to the west, but in Washington he was undoubtedly in close contact with Cass and Schoolcraft, both of whom were his close friends and both of whom also were well aware of the potential of the region. Cass, now a member of the federal administration, could not actively push the cause of Michigan, but he did work behind the scenes.[48]

Lyon soon realized that the boundary with Ohio was going to be determined purely on the basis of power politics and that Michigan was bound to lose. Resigning himself to reality, and with Cass's encouragement, he decided that "I for one shall go in for all the country Congress will give us west of the lakes."[49] If nothing else, he declared, "we can raise our own Indians in all time to come and supply ourselves with a little bear meat for delicacy."[50] Just a short time later he become even more emphatic: "My opinion is that within twenty years the addition here proposed [the western part of the upper peninsula] will be valued by Michigan at more than forty millions of dollars, and that even after ten years the State would not think of selling it for that sum."[51]

The same day he wrote in another letter that "a considerable tract of country between Lake Michigan and Lake Superior is known to be fertile and this with the fisheries on Lake Superior and the copper mines supposed to exist there may hereafter be worth to us many millions of dollars. I shall continue to urge it, under the present circumstances, as of the first importance."[52] Cass was also working for the proposal. He was familiar with the potential of the district both from his own earlier expedition and from the report of Houghton.

Michigan's other senator-elect and congressman, however, both remained adamantly against the idea. Protest meetings were held in Michigan, and one in Detroit resulted in a petition to Congress signed by many prominent citizens. It declared that the Upper Peninsula would always be a sterile wilderness, and challenged anyone's right to change Michigan's boundaries.[53] Because of his willingness to accept the settlement, Lyon was charged with selling Michigan out, which he vigorously denied.[54] The *Detroit Free Press* editorialized that if Congress thought that "the people of Michigan can be reconciled to its provisions by extending their jurisdiction over the region of perpetual snows—the *Ultima Thule* of our national domain in the North—they are much mistaken."[55] At a public meeting in Monroe County, the citizens were explicit in their feelings: "We deny the power of any tribunal on earth to compel us to accept Jurisdiction over any portion of territory which we may not be willing to receive."[56]

For Lyon there were other worries. Instead of using the Montreal River at the far western end of the Upper Peninsula as the western boundary of the state, there were proposals to use the Ontonagon or even the Chocolate [Chocolay] River, which would have left the entire mineral district in the Wisconsin Territory. Gradually, however, a more realistic view emerged. It became increasingly clear to all that political reality would dictate in the end, and that the Ohio strip was gone forever. In the final deliberations, the northwest boundary was almost forgotten in the rush to get the whole thing over with, and on January 26, 1837, Michigan officially entered the Union in two pieces. In one of them, almost as an afterthought, was the copper district.

And where, during all this, was Houghton? He, after all, knew the district best. His work of 1831 and 1832 had served to draw renewed attention to the copper and to reawaken interest in its potential. He had become a prominent and respected member of the community, with contacts at the highest levels of government. He was widely known for his interest and competence in natural history, particularly geology, and was at that very time laying the foundations for the first geological survey of Michigan. Shortly, he was to become the first state geologist. In 1831 he had written that the question of the amount of ore that the Keweenaw could produce could be settled only by "minute and laborious examination," which was subsequently to constitute perhaps his major accomplishment.

And yet, Houghton does not seem to have played any significant role in the events that made the district a part of Michigan. Governor Stevens T. Mason may have consulted with him as he sought to formulate his position on the struggle with Ohio,[57] but there is little evidence that Houghton played any role in the intense public debate over the question, even though his opinion would have undoubtedly carried great weight.[58] Recently, in a published article on Houghton's life, an author has claimed that the plan to include the copper district in Michigan was in fact "Houghton's proposal," and that the acceptance of the compromise by the citizens of Michigan was largely due to his influence.[59] Very little evidence was offered to support this claim, however, and the case presented does not seem convincing. In any event, whatever the exact details, the result was that the copper region now did fall within the boundaries of the new state of Michigan, and therefore it became a part of Houghton's new responsibility in the position that he was about to accept.

The State Geologist

When Michigan was admitted to the Union the state geological survey movement was well under way throughout the country. This distinctive American phenomenon represented one of the earliest efforts by governmental units in the United States to give official financial support to scientific investigations, always, to be sure, with a view toward practical results.[60] The direct forerunner of the state surveys was the private geological survey of Rensselaer County, New York, in 1821. It had been conducted by Amos Eaton and financed by Stephen Van Rensselaer. The first individual state surveys followed soon, beginning in the South with the Carolinas. Gradually the idea crystallized that a state should support geological research in looking out for its economic interests, and this provided opportunities for geologists to carry out research without having to rely exclusively on private money.

This was a significant step forward in the relationship between governmental units and the advance of science, but it was one that also had within it the seeds of possible conflicts of interest. Because surveys cost money and state legislatures were not inclined to frivolous spending, successful state geologists were often required to be able to argue their case persuasively, and friends in high places were very advantageous. In times of prosperity surveys flourished, but in hard economic times they were early candidates for curtailment or elimination. However, the movement grew, and in the 1830s and 1840s twenty states established surveys. By 1845 they covered most of what was then the United States.

To a degree a state geologist was in a difficult position. On the one hand, the state was invariably looking for practical results, for information that would be of immediate value and would promote economic development and attract new settlers. On the other hand, many geologists, seeing themselves as emerging professionals, were interested in their subject from a more strictly scientific perspective. Many conflicts resulted, and the middle road was not always easily found. Edward Hitchcock in Massachusetts and David Dale Owen in Indiana made explicit efforts to give their legislatures the practical results they were seeking before going on to "pure" science.[61] James Hall, the long-term state geologist of New York, successfully took the opposite approach. A very persuasive man, he laid the foundations of American paleontology in volumes of expensive books loaded with costly engravings of fossils, while managing to keep the appropriations coming by the force of his personality. Perhaps William Mather, also on the New York survey, spoke for a more generally accepted middle position: "Details and facts, belonging strictly to pure scientific geology, will not be made public until the final report. The object of the annual reports is to give publicity to such facts and localities as may be of practical utility, so that the benefit may be derived from a knowledge of them during the progress of the survey."[62] This approach, as subsequent events demonstrated, coincided almost exactly with Houghton's interpretation of his position as state geologist of Michigan.

The newly formed State of Michigan was acting, then, within a broad and developing tradition when it quickly authorized a geological survey. On February 23, 1837, less than a month after statehood, Governor Mason signed the survey act into law.[63] Undoubtedly the possibility of practical, immediate benefits must have been uppermost in the minds of those favoring the survey. Some writers have claimed that one reason the survey was instituted was to counteract rumors circulating in the East that Michigan was an unhealthy land of poor soils.[64] This seems unlikely, however. Michigan had been the object of such rumors, but at a much earlier period. In the immediate prestatehood period Michigan was, in fact, the most popular destination for pioneers heading westward, and during the 1830s it had the largest percentage increase in population of any state or territory.[65]

A more likely motive was the anticipated development of the state's natural resources. At least at the beginning, copper does not seem to have played a part; the dispute over the western part of the Upper Peninsula was barely over. Instead, according to Michigan's second state geologist, the overriding original motive was the desire to ensure an adequate supply of salt, an essential in the pioneer economy of that day. A great amount of Houghton's time and effort was indeed devoted to salt wells in his first years as state geologist.[66]

The degree to which the plan for the state geological survey was Houghton's own is not entirely clear, but certainly his contribution must have been substantial. According to one source, the plan and its successful implementation was entirely the personal doing of Houghton. Despite opposition from frugal legislators, he used his personal tact, persuasive powers, and friendship with the governor and others in high places, so that "over every impediment thrown in his path he triumphed."[67] He was evidently willing to leave his medical profession behind and return, at least for a time, to his love of nature and to public service. His appointment as state geologist, a position the governor declared to be one of the most important in Michigan government, was hailed with universal satisfaction.[68]

The first three years of the survey, 1837-39, were devoted primarily to developing the salt springs and to the general geology of Michigan, particularly the southern peninsula and the eastern portion of the Upper Peninsula. That work soon received favorable comment. In a review of state geological surveys in 1838, the *American Journal of Science* explicitly commended Michigan: "This young State has set a laudable example, in ordering a geological survey, under Dr. Douglass Houghton, which he has carried on with peculiar zeal," and it numbered him among "a phalanx of explorers in geology and natural history . . . respectable for knowledge, zeal, perseverance and success."[69] Professionalization was clearly proceeding rapidly, and Houghton was at the heart of the process.

A fair summary of his first annual report was also included in the journal, and the hope was expressed that the needed additional finances would be forthcoming from the state. As part of his early efforts, Houghton visited salt works in several other states, but, despite some promise in several attempts at well drilling, profitable salt production was for a variety of reasons not achieved in Michigan. These years were also marked by a reorganization of the structure of the survey to include departments of botany, zoology, and topography, all under the state geologist, by financial crises, and by calls for justification of the moneys appropriated by the legislature.

Even before the first year was up, the governor, perhaps as a result of pressure from other officials, called on Houghton to report on the "progress and advantages" of the survey. Houghton could not conceal entirely a note of irritation at having to justify the survey so soon.[70] In 1838 he was called on to respond directly to the House of Representatives about the "direct benefits" that could be anticipated by agricultural interests in the state from the survey.[71] Again, in 1839, with the situation exacerbated by a severe financial panic, he was invited to respond to the Senate "with all convenient dispatch" as to "what loss, injury or detriment might result from a temporary suspension of said geological survey." Houghton responded with a pointed and vigorous defense of all aspects of the survey and particularly of his assistants, who were "men who have left lucrative professions," and "who would immediately find an active field of labor, were they to be separated from this work, and who, in all human probability, could never be induced again to undertake the task in which they are now engaged."[72] Nevertheless, December of that year found him again complaining of the effects of financial cutbacks on the survey's progress. It was a battle he had to fight continuously, and he was forced eventually to agree to terminating the departments of botany and zoology.[73]

From early in its beginnings one of the goals of the survey had been to establish a central repository at the University of Michigan for the large collections of all types that would accumulate as a result of the investigations, and it was during this period that Houghton began his association with the university. In his initial defense of the survey, he had listed "furnishing our university with specimens of Natural History" as one of its three major advantages.[74] In the fall of 1839, on top of his duties as state geologist, he was appointed professor of geology, mineralogy, and chemistry, effectively becoming the founder of those departments.

The university at that time was relocating from Detroit to Ann Arbor, and actual classes awaited the completion of appropriate buildings. Shortly thereafter he wrote to John Torrey, his botanist friend and head of the biological portion of the New York survey, that "I shall commence removing my Geological, Mineralogical, and Zoological collections to Ann Arbor within a few weeks." He did not anticipate living in the city until the university began holding regular classes, which was expected to be some time yet in the future.[75] At the time, completing the survey was still his major goal.

As for the survey, by the end of 1839 the "preliminaries" were out of the way. The general geology of the Lower Peninsula and the eastern half of the Upper Peninsula, which, structurally speaking, is relatively simple, had been defined and outlined. There remained only the Upper Peninsula's remote, western half, whose geology was far more complex and which, not incidentally, contained the copper district. Houghton recognized the importance and difficulty of the work that yet remained, but he looked forward to it with confidence. The deadline for his final report, the ultimate objective and capstone of the entire survey, was March 1, 1841. He believed that this deadline was realistic, and that the conclusion of the survey was in sight. He confidently closed his third annual report, covering the work of 1839 and dated February 3, 1840, with these words:

> The geology of that district extending from Keweena [sic] point to, and including the Porcupine mountains, and stretching far into the interior, will require much minute examination, for it is within this district that the rocks containing the copper ore of lake Superior [sic], are embraced. Were it not that I have already examined this country sufficient to know to what point to direct particular attention, it would be impossible to accomplish a work, involving such immense labor and hardship, within the time specified by the act of organization; but as it is, aided by the efficient and industrious assistants connected with the department, I can safely say, that the whole will be accomplished within that time.[76]

But not for the last time would Houghton's estimate of how long he would take to complete the survey prove to be quite unrealistic.

5

THE COPPER REPORT AND THE COPPER RUSH

The year 1840 was momentous in the history of Keweenaw copper. It marked the beginning of a decade that saw greater advances in the understanding and development of the district than any corresponding interval before or since. Between 1840 and 1850 the district was transformed from a remote and rumored region of suspected but unconfirmed potential into a working, producing mining district whose uniqueness was well on the way to being understood. Without any doubt it was the work of Houghton and his assistants during the field season of 1840 that initiated those changes. The results of that season were compiled as Houghton's fourth annual report, and of his seven annual reports that covered 1837 through 1843, this was by far the most important. It might be fairly characterized as the single most influential document in the history of the district.[1]

Its importance extended beyond geology. In a sense it provided the basis for the social foundations of the region as well. In 1840 the south shore of the lake was still mostly a wilderness with a tiny, largely transient population of Indians and explorers. Small settlements existed at the Sault and at La-Pointe in the Apostle Islands, but the coastline between was the permanent home of only a few hardy missionaries and traders. By 1850 much had changed. Towns had been established at Copper Harbor and Ontonagon. Roads were being built, local trade was beginning, mining efforts were in progress, and the establishment of an indigenous American society was well under way.

Before the beginning of that important field season of 1840, Houghton and his chief assistant, Bela Hubbard, journeyed to Philadelphia to be part of the first organized meeting of the Association of American Geologists and Naturalists. In 1838 a group of New York geologists had proposed that steps be taken toward forming a national geological organization. Holding regular meetings was one of their major objectives so that geologists from around the country, and particularly those in the state surveys, could get to know one another personally, establish mutual confidences, and discuss items of mutual interest and concern. The association that emerged became one of the most important and enduring scientific organizations in the country, and in 1847 it metamorphosed into the American Association for the Advancement of Science. Its formation was an important step in the professionalization of geology, and of science in general, in the United States.[2]

In Philadelphia, Houghton and Hubbard joined some of the most prominent geologists of the country for a three-day meeting. The presence of the Michigan geologists has been specifically cited as the evidence that "demonstrated that a truly national society was possible." The intention of the organization to be restrictive in its membership was explicitly stated: "Resolved, that no person shall be considered as qualified to become a member of this Association who is not devoted to Geological research with scientific views and objects."

In subsequent years the meetings were sometimes the scene of intense and heated debate over points of theoretical difference. However, men like Benjamin Silliman and Edward Hitchcock, the state geologist of Massachusetts, took a paternal interest in maintaining a balance between the potentially conflicting influences within the group. In 1842 Charles Lyell, one of the founders of modern geology, took an active role in the annual meeting during his tour of North America, thereby adding greatly to the prestige of the organization. Perhaps William Rogers of the Virginia survey spoke for most when he said, "For us such reunions of the scientific brethren as our Association of Geologists are of precious value and form the best compensation we can enjoy [sic]." The professionalization of geology was well under way, and Houghton was in on the ground floor.[3]

For him, the subsequent annual meetings were a major source of information about the geology of other regions of the country. It also gave him a forum for disseminating information about Michigan copper, and he played a continuing and active role in the life of the association. After the first meeting ended, Houghton passed through New York on his return to Michigan, where he learned of that legislature's intent to extend the New York survey for two years. This provided him with a precedent when he decided he needed more time to complete his Michigan work after the field season that fall.[4]

The 1840 Field Season

Back in Michigan, he planned the summer's work. Bela Hubbard and Columbus C. Douglass, Houghton's cousin, were to be his principal assistants. Hubbard was another New Yorker who had headed west to Michigan to seek his fortune. He had, in close association with Houghton, become deeply involved in real estate speculation, which provided most of his income during his Michigan years. In 1837 Houghton hired Hubbard and Douglass into the survey, Hubbard being a capable geologist in his own right. Together, they found that information gleaned on survey trips through the state could provide important clues for possible profitable real estate and mining investments.[5] Also making the journey were Bela's brother Fredrick and Charles W. Penny, young men of the Houghton circle getting a taste of life in the wilds of the north. Along with Houghton, Hubbard and Penny kept journals of the trip, which document very well the events of this important expedition.

They left Detroit late in May and assembled at the Sault for the trip west. The party, now with the addition of voyageurs and woodsmen but temporarily minus Houghton, who remained behind for several days, headed out along the southern shoreline of the lake on the first of June. Camped for four days at Grand Marais while awaiting Houghton, they spent their time marveling at the scenes of the Grand Sable with its huge dunes and steep banks of sand that towered above the lake. Finally, on the eleventh of June their leader arrived and they set out together toward the mineral district.

Moving on westward, they gazed at the cliffs of the Pictured Rocks as they passed, noting faithfully the sandstone that made up so much of the southern coast of the lake. The real geological work began when they reached the Chocolate [Chocolay] River. This river had long been known to mark a fundamental change in the physical and geological characteristics of the land. To the east, extending all the way to the Sault, the land was relatively low and geologically simple. Wherever the bedrock was not covered by overlying surface deposits, nearly flat-lying sandstones were exhibiting along the coast. Westward from this river the land was more rugged, the ancient rocks were often folded and deformed, and the geology generally was far more complex. "We now come into the region of the primary rocks," noted Hubbard, "where more laborious work attends us geologists."[6]

At this time, the Chocolate River also marked the beginning of the Indian lands. These lands had not yet been ceded to the United States, although the treaty of 1826 did allow for exploration and the right to search for and carry away any metals or minerals from any part of the country. Near the present city of Marquette, Houghton examined and mapped the rocks along the coast in some detail. The travelers were kept busy collecting many samples that required labeling and storing, and before they were even close to the copper region they had already filled over six barrels, each sample being carefully wrapped in paper and labeled.[7]

Skirting Keweenaw Bay, they finally headed up the east side of the peninsula. On the southeast side of Keweenaw Point they had their first encounter with copper-bearing minerals in the form a few small veins of black and green copper compounds near the Montreal (or Cascade) River. Houghton doubted if they had much value, but a few miles beyond he found a more promising location, where they collected samples for further analysis. Rounding the point, they arrived at Copper Harbor on the third of July. Remaining until the eighth, they blasted and examined closely the famed "green rock." Six casks of samples were packed, but they were left to be picked up on the return trip. Houghton's opinion was that this vein could undoubtedly be worked advantageously, and as they investigated the area further, Penny observed of Houghton that, "as he becomes better acquainted with the geology of the Point, his confidence in the copper deposits increases."[8] There in the wilds they spent a pleasant fourth of July, setting off a particularly large blast in celebration and feasting "in the most sumptuous manner" (fig. 13).

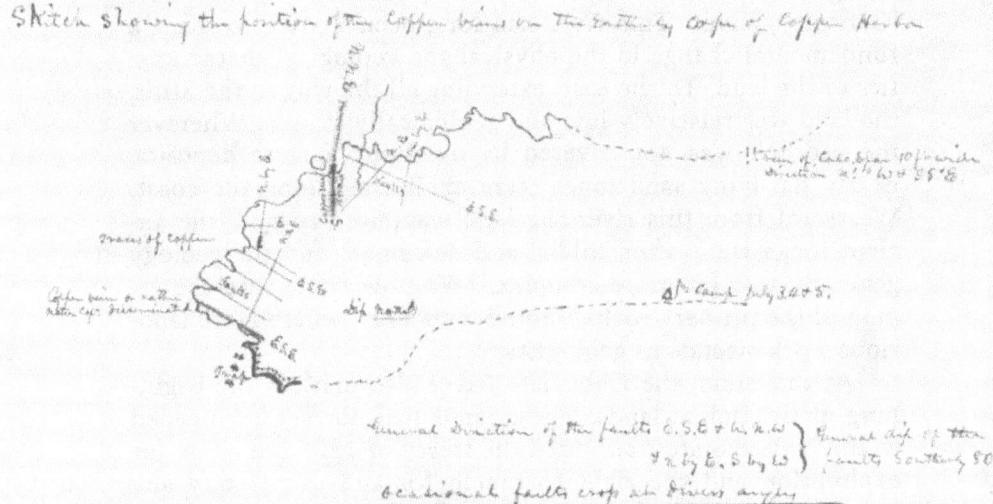

Fig. 13. Houghton's sketch of the green rock location on the east cape of Copper Harbor, as drawn in his 1840 field notes. North is to the top. Their campsite is shown by a tent and flag. The green rock vein is labeled "Copper vein 15 feet thick," and it can still be seen today as a broad white and green stripe extending into the lake. Courtesy Michigan Historical Collections, Bentley Historical Library, The University of Michigan.

Continuing westward, the party made its way along the coast heading for the Ontonagon River and the now obligatory visit to the boulder. Reaching the river on the tenth, they were greeted by Chief Buffalo and a band of Indians occupying the lodges on the traditional village site on the west bank. Anxious to proceed upstream, the party got under way despite indications that the Indians were not pleased by their presence. The Indians pursued them upriver and tried to blackmail them by demanding that they make payment in the form of supplies. The bargaining went on for an hour, but Houghton was very firm, after which the Indians retreated and the explorers continued. Because of the rapids, some of their supplies had to be left behind with four men as guards. Houghton gave them "the strictest orders to fire upon the first indian [sic] who touched the baggage,"[9] and threatened that anyone not obeying would forfeit his own life.[10] They subsequently learned that the Indians had heard of their coming long before they arrived and had resolved to get from them what they could, but no further problems were encountered.

They reached that "far famed mineral wonder" on Sunday, July 12. The exertions of the trip led Penny to put it succinctly: "Whoever visits the rock pays well for his curiosity."[11] They found broken chisels and a partly detached piece exactly as Houghton had left them during his previous visit nine years before.[12] Houghton worked them hard, for he was anxious to get some sizable specimens for the university collections, and they managed to hack off some twenty-five pounds worth as Houghton took careful notes (fig. 14). Even so, it was very difficult, as all the easily removed projections had long before been carried away. On the return toward the mouth they again met some Indians who told them that "the Buffalo" had gone upstream to find out how much they had cut from the boulder, but when they reached the lake again the village was deserted. "Although well satisfied with our trip," Penny dryly observed, "we are particularly gratified to know that we shall not have to perform it again."[13]

Coasting farther westward, and after meeting only a few traders and employees of the fur companies, they finally reached LaPointe, for them a welcome return to civilization. This settlement, on Madeline Island of the Apostle group in the Wisconsin Territory, had been for many years the focus of the entire western Lake Superior region, both for the Indians and the fur companies. The American Fur Company had built many business structures and houses, and it was the site of both Protestant and Catholic missions. The rough appearance of the travelers, after so long in the woods, quickly earned them each Indian names. Penny, his thick black beard amazing the Indians, became "the Bear." Hubbard was "Red Wolf," and Houghton, retaining his name from his earlier visit, was "Mus Ke Ke we nin ne," the "Medicine Man."[14] With a shave and a change of clothes, however, they quickly adapted again to the sound of the church bell, the music of the pianoforte and violin, games of chess, and the company of "the more gentle sex."

On the twentieth of July, the expedition split up. Houghton and some field men remained to carry on more detailed work; Hubbard and Douglass returned to continue their survey of the Lower Peninsula. At LaPointe, Penny had become somewhat smitten by "the prettiest Indian girl I ever saw," whose "figure was much superior to that of ordinary European

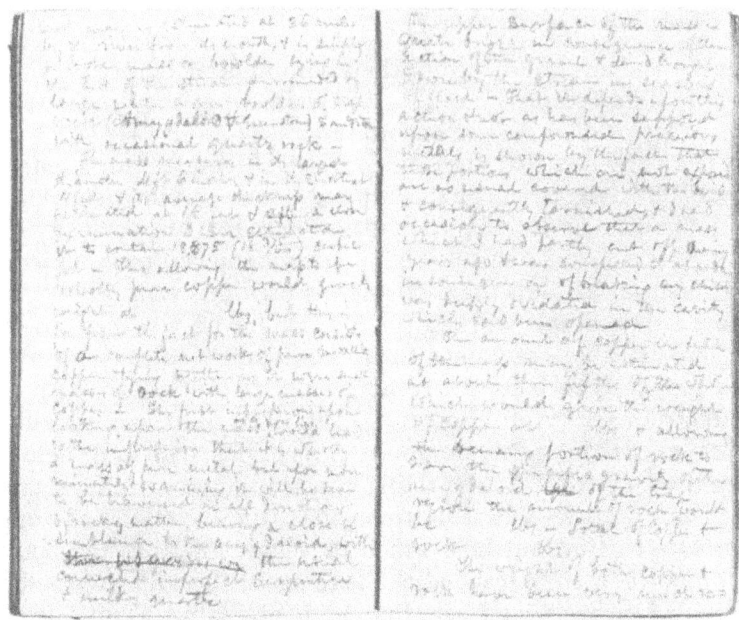

Fig. 14. Houghton's field notes describing the Ontonagon Boulder. In 1972 Houghton's long-lost original field notes turned up in a box of miscellaneous Michigan papers being auctioned in New York, and they were then acquired by the Clarke Historical Library. The pages shown here were written July 12, 1840, and describe the Ontonagon Boulder on what was Houghton's third visit to it. The blanks were left for weights that he intended to calculate later. Courtesy Clarke Historical Library, Central Michigan University.

[The mass is at a distance...] that may be estimated at 26 miles by the river from its mouth, & is simply a loose mass or bowdler lying in the bed of the stream surrounded by large water worn bowlders of trap rock (Amygdaloid & greenstone) sandstone with occasional quartz rock -

The mass measures in its largest diameter 4 ft 6 inches & in its shortest 4 feet & its average thickness may estimated at 1 1/2 feet & after a close examination I have estimated it to contain 18.375 (18.375/1000) cubic feet - This allowing the mass to be wholly pure copper would give its weight at   lbs., but this is far from the fact for this mass consists of a complete network of pure metallic copper tying together as it were small masses of rock with large masses of copper - The first impression upon looking upon this mass would lead to the impression that it is wholly a mass of pure metal but upon more minutely examining it will be seen to be transversed in all directions by rocky matter, bearing a close re-semblance to the amygdaloid, with ---------------------- the usual connected imperfect serpentine & milky quartz.

The upper surface of the mass is quite bright in consequence of the action of the gravel & sand brought down by the stream in season's of flood - That it depends upon this action & not as has been supposed upon some compounded precious metals is shown by the fact that those portions which are not exposed are as usual covered with the oxide & consequently tarnished, & I had occasion to observe that a mass which I had partly cut off many years ago & was compelled to abandon in consequence of breaking my chisels was deeply oxidated in the cavity which had been opened.

The amount of copper in bulk of the mass may be estimated at about three fifths of the whole which would give the weight of copper as   lbs & allowing the remaining portion of rock to have the specific gravity of the amygdaloid --- of the trap region the amount of rock would be   lbs - total of copper & rock   lbs.

The weight of both copper & rock have been very much re-[duced since the mass was first visited ...]

ladies." In response to her pleas, he had invited her to accompany them on their journey back. Unfortunately, this plan was vetoed by the majority of the party and he, feeling badly about going back on his word, was forced to leave her behind.[15] They retraced the long route back to the Sault without incident, and on the thirty-first, after exactly two months on the lake, they arrived safely back at the rapids.

Houghton's plan was to continue his investigations on Lake Superior for the remainder of the season, including a visit to Isle Royale. On the twenty-second, he wrote a long letter to Governor Woodbridge, summarizing his progress to that point. On July twenty-eighth, he and three men left La Pointe for the island on the schooner *Wm. Brewster*. Once there, officials of the American Fur Company provided him with a boat, and went out of their way to assist him in his travels. He circumnavigated the island, and remarked on the similarity of its geology to that of Keweenaw Point, but the few small veins of native copper he found did not strike him as being of any practical importance. A few brief observations were also made on the mainland to the northwest of Isle Royale, and the eighth of August found them back at La Pointe again.

Two weeks were spent near the western limits of Michigan. They concentrated particularly on the country around the Montreal River where, the next year, he was to spend much of the field season trying to maintain Michigan's claims in the region of the disputed boundary with Wisconsin. While camping in this region Houghton, although wise in the ways of the woods, was fascinated by the technique his woodsmen used one evening to grill a porcupine. One of them, using a method learned from the Indians, inflated the animal by inserting his pipe into its anus and blowing into it. After being cooked in that state the quills were then easily scraped off. After watching the procedure, Houghton wrote in his notes that "the perfect nonchalance with which the voyageur took his pipe from the rectum and placed it in his mouth was beyond conception."[16]

Finishing their investigations, he prepared to bring the summer's efforts to a conclusion. On the twenty-fourth of August they left La Pointe and began working their way along the southern coast of the lake, retracing their earlier steps back toward the Sault. At Eagle River, where five years later he was to meet his death, he made a detailed examination of several copper-bearing veins. Rounding the point and proceeding down its eastern side, they entered Portage Lake, where Houghton collected enough information to make a map and geological cross-section of the peninsula (fig. 15). On the fifteenth of September they arrived back at the Sault, and on the twenty-fourth he arrived at Detroit on board the steamboat *Missouri*, bringing his most important single field season to a close.

Theoretically, with the completion of the 1840 field work, the data-gathering phase of the state survey should have been completed and the final report should have soon followed as the survey's finishing and culminating document. At the end of his annual report for 1839 Houghton had expressed confidence that the survey would be completed within the originally allotted time. Yet, it was now becoming increasingly clear to him that the long-anticipated final report was not likely to appear in the immediate future. On the prospects of finishing the survey soon, Houghton now reported to the governor that "with the utmost of my powers I have only been able to accomplish a portion of this arduous and difficult task."[17] He pointed with pride to what had already been done, noting that Michigan was vastly more difficult to explore than New York, where the survey had more than four times the personnel of that of Michigan and even there the original four-year term had been extended for two extra years. Surely, he said, the state would not want an imperfect survey, and "I would be unwilling to involve my reputation in connexion [sic] with these surveys while portions of the work remain incomplete."

Houghton's observations in this regard undoubtedly were valid and reflected very real concerns on his part. However, experience in other states had also shown that completing any state geological survey invariably required a realistic compromise between the conflicting desire for completeness and scientific accuracy on the one hand and the need to produce a document of value to the residents of the state on the other. In the case of Michigan, this hesitation marked the beginning of a series of delays that was to eventually encompass the next five years, and the result was that the state survey was never completed and the final report was never written.

It is difficult not to wonder at the reasons for the extreme delay on Houghton's part in completing the work. Even granting fully the very real problems he encountered in obtaining adequate financing and in retaining the needed assistants, the entire picture would seem to be more complex. Houghton was

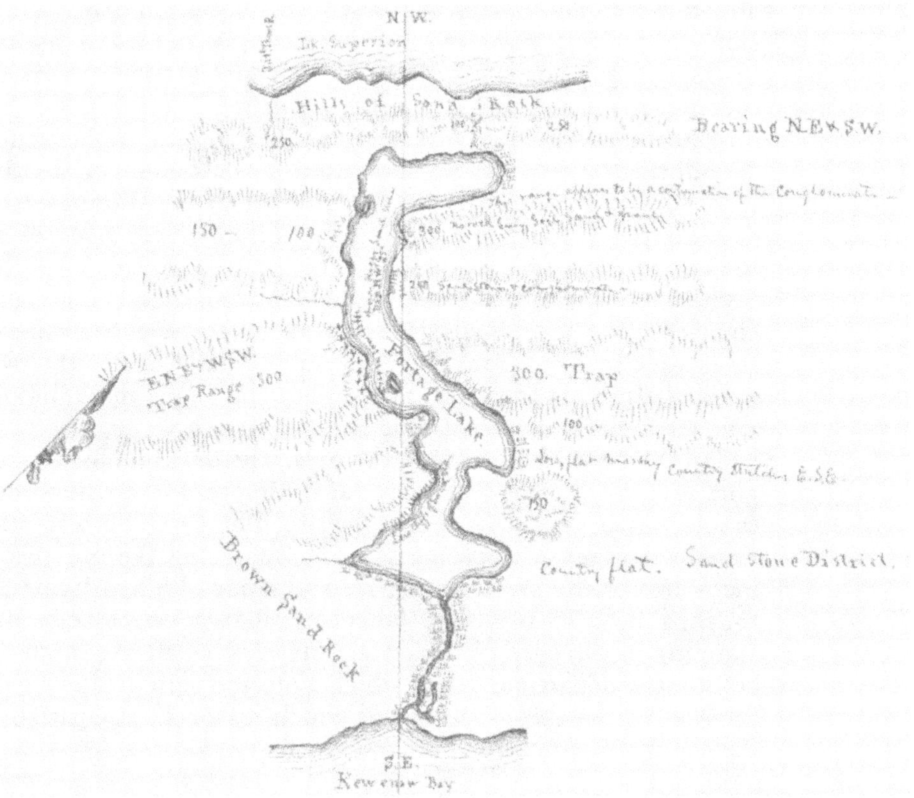

Fig. 15. Houghton's map and cross-section of the Portage Lake region, as included with his handwritten transcription of his field notes. The flat-lying sandstone to the southeast and the tilted traprock and conglomerate layers to the northwest are clearly shown. Courtesy Michigan Historical Collections, Bentley Historical Library, The University of Michigan.

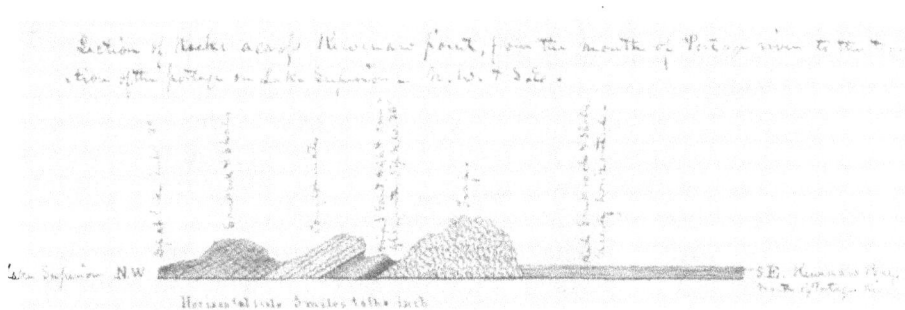

a man driven and drawn by many forces acting in many directions. These forces were to bear increasingly on him in the following years, and he evidently did not always find it easy to organize his sometimes conflicting priorities.

On the seventeenth of December, responding to Governor Woodbridge's request, he submitted a "hasty report" in the form of a letter summarizing his findings of the summer. In a cover letter, Houghton regretted that his chemical analyses of the copper ores had not yet been completed. Nevertheless, he invited Woodbridge to examine them, noting that "I am busily engaged upon their analysis. I hope soon to be able to know their precise *condition*."[18] Yet, a month later and only two weeks before his fourth annual report, the single most important geological document that he ever was to write, was due, he still had not finished with the ores. The deadline for the report was the first of February. According to a letter he wrote to his father only two weeks before that date, he was still "busily engaged for several days" in analyzing the ore specimens from Lake Superior. As for the report, he had "scarcely touched it." He was not, however, unduly concerned. His standing in the eyes of the legislature was, as always, high: "I believe the department over which I preside is in greater favor than it ever has been before, for which reason I anticipate but little embarrassment."[19]

The Copper Report

He made the deadline. The fourth annual report, dated February 1, 1841, and known thereafter simply as "the copper report," became the founding document of the subsequent copper rush. Although mostly concerned with the Upper Peninsula, it was not limited to the copper district, and it also included appended reports by his assistants C. C. Douglass, Bela Hubbard (both largely on the Lower Peninsula), Fredrick Hubbard, and S. W. Higgins, the topographer. The copper region was paramount in Houghton's work, however, and over one-third of his portion was taken up with a detailed discussion of the "Mineral veins." This is the most extended analysis of the mineral region that Houghton ever wrote, and it provides the best available treatment of his understanding of the nature and structure of the Michigan copper district.[20]

Earlier, he had written to the governor that he had remained in the field "until the severe fatigue and hardships to which I had been exposed had so far impaired my health as to render me wholly unfit for duty," and he noted his constant exposure to the "wet, cold, and the vicissitudes of the weather."[21] In the report, he again expressed his regret that "the sufferings and hardships to which I have been exposed in conducting the field work over the wilderness portions of our state, have so far impaired my health, as to render it impossible for me to enter into so minute details as has been anticipated." He also bemoaned the gross inaccuracies of the maps generally available at the time, noting that many features were depicted "as far from the truth as could be conceived."

After a general geographic description, he grouped the rocks of the northern portion of the Upper Peninsula into nine units, ranging from the oldest, primary rocks found just west of the Chocolate [Chocolay] River to the "Tertiary Clays and Sands," which would now be called the glacial deposits of the Ice Age, that overlay all the others. The most important of these units, from the perspective of the copper district, were the traprocks and overlying conglomerate and sandstone.

Houghton clearly defined the basic structure of the district and he recognized, both on the Keweenaw peninsula and on Isle Royale, the traprock unit now called the Portage Lake Volcanics and the overlying conglomerates and trap. Within the trap, Houghton also distinguished between the more massive, compact portion to the southeast and the more amygdaloidal portion to the northwest. There is more to be said about his understanding of the geology of the Keweenaw, but there is also no question that this work of Houghton became the basic starting point for all subsequent studies.

The major significance of the work of 1840, however, did not lie in the analysis of the general geology of the structure and nature of the rocks, but in its implications for the future development of the mineral riches of the Keweenaw peninsula. Even before his annual report for the year was compiled, Houghton discussed some of the results of the field season in several letters. That December, he supplemented the letter he had sent to Governor Woodbridge the previous July with another, as a "somewhat informal" report of his efforts. His words on the mineral district were undoubtedly music to the governor's ears. Its development would, he said

> not only add very much to the future settlement and prosperity of the upper peninsula, but also the state as a whole. Upon the whole (while I would carefully avoid exciting any unfounded ex-

pectations among our citizens, and caution them to avoid engaging in wild schemes with a view to gain sudden wealth) the examinations and surveys which have been made would serve fully to justify the conclusion that this region of the country will prove a continued source of wealth to our State.[22]

A few days later he responded to an inquiry about the district from Augustus Porter, a congressman from Detroit, with another letter noting that his copper samples from Lake Superior had "turned the heads of most of those who have examined them."[23] He also included a statement that was to become widely quoted, and which helped fuel the coming copper rush:

> In opening a vein with a single blast, I threw out nearly two tons of ore, and with this were many masses of native copper, from the most minute specks to about forty pounds in weight, which was the largest mass I obtained from that vein. Ores of silver occasionally occur with the copper, and in opening one vein small specks of native silver were observed.

"There can scarcely be a shadow of a doubt," he said, that the district "will eventually prove of great value to our citizens and to the nation. I hope to see the day when instead of importing the whole of the immense amount of copper and brass used in our country, we may become exporters of both."

It did not take long for the excitement to begin in earnest, for the copper district had long hovered in the background of the national consciousness. In 1839 a visitor to the district was already planning to begin mining with men imported from London, and such isolated visionaries had never given up on the dream of immensely productive copper mines.[24] Now, with a report from a respected geological professional, the excitement soon burst out. Houghton was always very careful to couch his conclusions in cautious, reserved language, and he repeatedly expressed the fear that the public could easily be deceived in mining schemes that had no possibility of success. Yet, as far as caution goes, he might as well have been talking to the wind. The initial impetus for what followed seems clearly to have been his copper report, cautiously worded though it may have been.

One last barrier to a full-fledged mining boom remained. The federal government still had effective control over the possibility of mining in the district, because the treaty of 1826 did not provide for mining by private individuals. Now, with the publication of the copper report, public pressure began to build to the point that, after some hesitation, it effectively forced a capitulation by the government to private enterprise. The year 1841 saw the beginning of the "Lake Superior Copper Fever," which, continuing until 1847, created the first great mining "rush" of American history.[25]

The news spread fast. The *Detroit Daily Advertiser*, in May 1841, probably expressed the sentiments of many: "Dr. Houghton has submitted a mass of valuable and interesting matter which cannot but excite the earnest attention of every class of readers." Also included was his statement about the two tons of ore and the forty pound chunk of native copper.[26] The copper report, or parts of it, soon appeared in newspapers, journals, and in "handbooks" that appeared for the potential mining barons that began to gravitate toward the shores of Superior. One author has written, "as an inventor is associated with his machine, or a Congressman with a bill, from the beginning of the rush Douglass Houghton had been identified with the copper contagion."[27]

About this time Houghton also began to assume an increasingly greater role on the national geological scene. The second meeting of the Association of American Geologists was held in April 1841, again in Philadelphia. While in the East Houghton bought some chemical and geological books and visited the trap region of New Jersey, which he found to be "a miniature representation of that of Lake Superior, with the associated minerals bearing a like proportion."[28] At the meeting he participated fully and took the opportunity to address the group on the copper district. He gave a detailed synopsis of his report, and the summary of his remarks took up three pages of the next issue of the *American Journal of Science*.[29] He noted the "dykes of trap" found between the lines of stratification of the sandstones and conglomerates, and the copper-bearing "true veins" that cut across those rocks. On the important question of the mining prospects, he was his usual cautious self, but concluded that "upon the whole they may be looked upon favorably rather than otherwise."

The pressure to exploit the copper continued to build, but the Indians still owned the land. Finally, in 1842 the Treaty of La Pointe was negotiated. Proclaimed in March 1843, it ceded to the United States a region that included all the land in Michigan west of the Chocolay River. The way was thereby cleared for the copper fever to rise to new heights, and interested persons soon began moving into the area. One of the first of these, however, was not interested in copper riches in any abstract sense. His goal, a very specific one, was the great copper boulder.

The Boulder and the Rush

The removal of the copper boulder from its resting place on the banks of the Ontonagon River had long been the dream of many, but all attempts had failed, defeated by its great weight and the remote and inaccessible location. It had become, however, a widely known symbol of the untamed wilderness, and carrying it away could perhaps represent man's triumph over nature. Henry Schoolcraft had not forgotten it since his visit on the expedition of 1820. When appointed to the first board of regents of the University of Michigan in 1837, he submitted a resolution asking the state geologist to obtain the boulder for the university, and he also suggested that the university could pay any transportation expenses involved.[30] The state apparently had better things to do with its money, and that effort was not made.

Success finally came from an unlikely source. Julius Eldred was a hardware merchant from Detroit who lived an uneventful life until he decided that fate had designated him to possess the famed copper boulder. He had heard the tales of earlier attempts to move it, but he was determined to succeed where the others had failed. He visited the site in 1841 at which time he bought the rock, so he thought, from the Indians for $150. He made a second visit in 1842 to prepare for his major assault of 1843. In the meantime, however, the treaty with the Indians had been ratified. Several exploring parties began moving into the district and, because of the long-standing reputation of the mass, the Ontonagon region was the focus of many of these efforts. When Eldred and his crew arrived back in the spring of 1843 he found that his was only one of several groups that claimed the right to possession of the copper rock. Undaunted, he bought it again, this time from a Colonel Hammond, for some fourteen hundred dollars. The rock was getting expensive.

He nevertheless set to work with a crew of up to twenty white men and Indians.[31] Carrying their equipment far inland, they built a "road" over four miles long through the rugged hilly forest to bypass the upper rapids of the river. With very great effort they raised the copper mass up the fifty-foot-high bluff by means of a capstan, after which the crew loaded it onto a small crude vehicle mounted on railroad wheels. Taking up the track from behind the car and re-laying it in front, they pushed and pulled the car up the hills and down the valleys over their rough trail, finally reaching the river below the major rapids. There they loaded the mass onto a raft, but the trip down to the lake hung in jeopardy because of the low water level of the river. A "providential shower" raised the river level just enough to carry the raft over the remaining rocks, and it arrived safely at the mouth.

Eldred's problems continued. The check he had used to buy the boulder the second time bounced, and Hammond was back demanding his property. Eldred, not about to give up now, managed to patch things up and Hammond accepted $1,765 as total reimbursement. Unfortunately, this was not the last of Eldred's difficulties. In June 1843 the government mineral agency had officially opened in Copper Harbor under Gen. Walter Cunningham. The harbor, at the most northerly spot on the Keweenaw, was already becoming a recognized staging ground for forays into the interior by the prospective mining barons swarming to the country. The settlement already consisted of "nine tents, averaging six to a tent, which makes quite a society."[32]

When Eldred stopped at Copper Harbor after chartering the schooner *Algonquin* to carry his treasure back to Detroit, Cunningham presented him with a letter he had just received from James Porter, the secretary of war. Porter had been informed by the secretary of the treasury that he desired "the transportation to this City [Washington] of a Mass of native copper found near the shores of Lake Superior, in order to its being placed in the National Institute . . . as the object is really one of great importance to the scientific world."[33] Porter therefore ordered Cunningham to seize the mass in the name of the United States, and authorized him to call out troops to enforce his authority, if necessary. Eldred was chagrined, and Cunningham, after collecting another one hundred dollars in transportation costs, generously allowed him to accompany the boulder to Detroit.

Home at last, Eldred placed the mass on exhibit, charging the public twenty-five cents for the chance to gaze on the fabled treasure of the Northwest. Schoolcraft was apparently one of the first to pay the fee to see what he had visited in nature over twenty years earlier. Eldred's joy soon turned to sor-row, however. The United States district attorney demanded possession of the mass, and after repeated threats Eldred, declaring his "unwillingness to enter into a controversy with the Government of the United States, which every citizen should deplore," yielded the rock to the federal government. It was taken to Washington via Buffalo and the Erie Canal, attracting great attention as it made its way eastward.

Eldred accompanied it as far as New York, but he failed to get a passage on the ship that was to take it on the last leg of its journey. He then proceeded overland to Washington, where he faithfully kept watch for the vessel. When it arrived, he oversaw the transportation of "this great natural curiosity, this extraordinary specimen of the mineral riches of our country, this splendid specimen of the mineral wealth of the Far West," to its resting place at the war department. Eldred continued to press his case, and an exhaustive investigation into all aspects of the entire affair, which included depositions by Houghton, Cass, and Woodbridge defending the importance of Eldred's efforts, resulted in a special act of Congress in 1847. As fair compensation for his efforts in acquiring the "siren of mining adventurers," Eldred was awarded $5,664.98 for an object that, after all the years of being hacked and gouged, was found to weigh just over thirty-seven hundred pounds, with a market value at the time of some six hundred dollars. Soon forgotton, the mass languished in the yard of the war department for years before it was finally awarded a place of honor in the Smithsonian Institution (fig. 16).

The publicity had its effect. By the end of 1843 over a hundred requests had been made for permits to explore the Keweenaw and the number was growing rapidly. By the end of 1845 the number had grown to over seven hundred, but the permits soon turned out to be the source of more problems than they solved.[34] In fact, the biggest single obstacle to the orderly development of the district proved to be the lease-permit system that had been quickly imposed on the region in a futile attempt to bring some coherence to the rush that developed.

Prospectors had begun to move into the district as a result of the acquisition of the land by the federal government in 1843. The government responded by simply transferring the permit system that had previously been used, rather unsuccessfully, in the lead district of the Mississippi valley. This transfer, later held to be illegal, allowed a person having a gov-

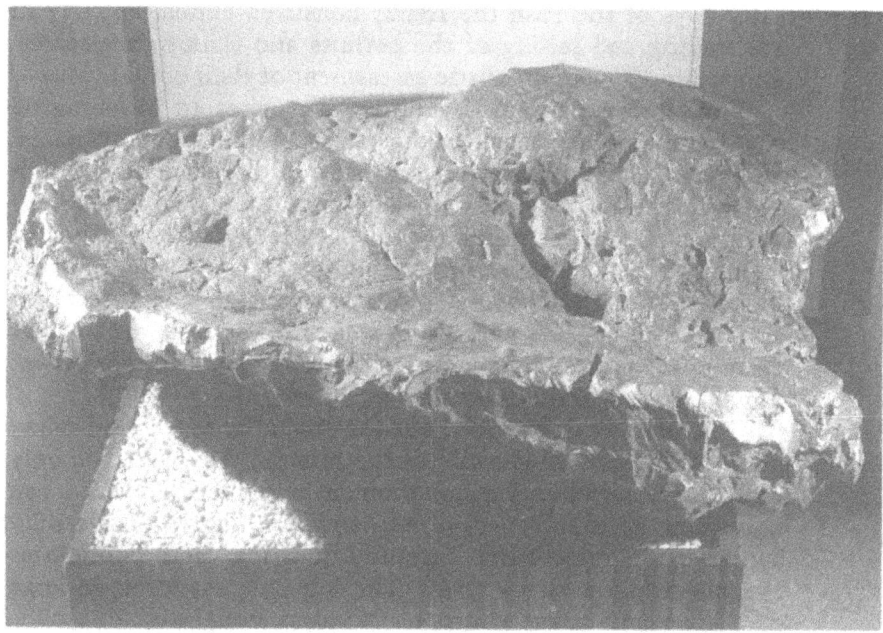

Fig. 16. The Ontonagon Boulder as it is displayed at the Smithsonian Institution in Washington, D.C. It is about four feet wide. Courtesy National Museum of Natural History.

ernment permit to register a claim for as much as nine square miles of land, a size later reduced to one square mile. The permits were free for the asking, but they required the bearer to file a claim within one year with the boundaries clearly marked, no easy task in a wilderness region that had not yet been surveyed. Once filed, the claim gave the bearer the "opportunity," on posting a bond of several thousand dollars, to lease the land for mining purposes for three years, renewable twice for a possible total of nine years. During mining, 6 percent of the metal was required as a tax; when the lease expired the land and, it appeared, any and all improvements, would then revert to the government.

It required no great intelligence to realize that in a region whose mineral values were still imperfectly known such a policy was virtually self-defeating. The chances of establishing a mining enterprise profitable enough to allow it to be handed back to the government with no compensation and after only a nine year period were essentially zero. As a result, during the early days of the rush the frenzy consisted almost entirely in the trading and selling of the permits and claims themselves, unrestrained by any realistic assessment of their actual value.

Corruption abounded. Political favorites in Washington got permits while others in the district found none available. Submitted claims mysteriously turned out to have been previously claimed by someone in Washington who had never been within a thousand miles of the Keweenaw. Those "lucky" enough to buy a claim from one of the many speculators who floated through the copper country and whose pockets seemed always to be bulging with them was also risky; "some of their permits are located where they can't touch land up, nor down, nor sideways—and others ain't nowhere."[35]

The entire system cried out for drastic revision, and many argued that an outright sale of the mineral lands was the only real American way. A major problem remained. The land had not been accurately surveyed, and without a universally agreed-upon system for carefully locating land boundaries chaos would inevitably result. The solution was to be, in large part, the result of the efforts of Douglass Houghton.

The publication of the copper report in early 1841 marked the high point of Houghton's contributions to the geology of Michigan. His reputation continued to soar, and his association with Keweenaw copper was widespread in the public mind. His following years, however, were not marked by any major geological discoveries. He continued working in the field and played a major role in settling the final boundary between Michigan's Upper Peninsula and Wisconsin, but his annual report for 1841 was only six pages long. He emphasized again that the field part of his work was largely complete, although "there still remains much to be done, in arranging the materials accumulated, for a final report upon the entire work."[36]

In 1842 most of his geological efforts were in the form of office work and his annual report was only four pages in length. By 1843 state funding had been reduced to a dribble, and his report was a brief three pages. In 1844, the geological survey had virtually ceased, although Houghton continued to be recognized as the state geologist. He attended the annual meeting of the association in Albany in 1843, where he participated in the discussions of geological topics, and his influence was clearly not diminished on the national scene. He was elected treasurer of the organization and was one of three members appointed to a special committee to "draw up a report for the next annual meeting, embracing all the known facts, bearing on the occurrence of native copper, in the trappean regions of the United States."[37]

During these years, other interests also began to occupy his time. The University of Michigan had spent, at the urging of Schoolcraft, four thousand dollars in 1838 to buy a major mineral collection from Baron Lederer in Austria.[38] Houghton began moving many of the innumerable specimens from his own collections to Ann Arbor in anticipation of taking up his duties as professor. A house had been built for him there, one of four built for the professors and "quite a splendid mansion," but he was reluctant to take up residence until the school was more thoroughly organized and there was "at least a sufficient number of the faculty to make a society of our own."[39] Houghton's other duties never gave him time to teach more than an occasional class, and he never moved to Ann Arbor, but his interest in the University and its welfare continued undiminished.[40]

During this period he was also very active in community affairs, being elected the first president of the Detroit Board of Education. To add to his burdens, he returned to Detroit in the spring of 1842, after an absence of three weeks, only to discover that "without any knowledge whatever, on my part, I have been elected mayor." He felt that he could not possibly carry out the duties, and declared that "I will probably resign after two or three weeks."[41] He nevertheless completed his term to general acclaim for his competence in administration, and higher political office looked increasingly like a real possibility. The importance of the final report for the geological survey may have begun to recede in his mind.

Still, the lure of the North remained. He felt the need for even more data from the mineral region on Lake Superior, but in financially difficult times the state was not about to subsidize further investigations. At the same time, the federal government surveyors were gradually pushing their linear surveys, by which the states were being divided up into townships and ranges, farther northward. Because these surveys, along with similar surveys throughout the country, divided the land into sections of one square mile each, this meant that a crew of men would pass within no more than one-half mile of every point in the region. Houghton had learned much of value from the linear surveyors who were working in the Lower Peninsula during his investigations there. He gradually developed the concept of a combined survey, one that would both subdivide the land and also make appropriate geological observations. In this way, and with little expense beyond that of the linear survey itself, he believed that a vast amount of geological data could be obtained at only a small fraction of the cost of a survey run for geological purposes alone.

At the meeting of the Association of American Geologists and Naturalists for 1844 held in Washington, Houghton read a paper entitled "Remarks on the Importance and Practicability of Connecting Geological Surveys with the Linear United States Surveys." He outlined the advantages of such a combined survey and cited similar practices in several European countries. His colleagues were so struck by the clarity and logic of his arguments that a resolution was adopted authorizing a committee, on which some of the most prominent geologists of the country agreed to serve, to "memorialize the proper department of the government, on the importance of connecting geological surveys with the surveys of the public lands."[42]

Governmental response was very favorable. The only real concern expressed was that the members of the survey crews might not be adequately trained in geology to make meaningful observations. Houghton resolved that concern in a most direct manner by agreeing to conduct the survey himself. In June 1844 he signed a contract with the federal government to "faithfully lay out and survey with reference to mines and minerals, four thousand miles of the township and sectional lines in that part of Michigan south of Lake Superior" at a cost of five dollars per mile. He confidently agreed to "bind ourselves, our heirs, executors, and administrators" to a performance bond of forty thousand dollars.[43] In his last annual report he wrote that he was determined to produce "more perfect geological and topographical maps of the upper peninsula than have ever been constructed of the same extent of territory in our United States."[44] He was on his way, back into the field again.

As his second in command Houghton chose William A. Burt, inventor of the solar compass, a good practical geologist, and a veteran surveyor who had been working on the linear surveys in Michigan for some time. It was Burt's party that, in the fall of 1844, first discovered significant quantities of iron ore in the interior region west of the Chocolay River.[45]

Houghton laid plans to continue the combined surveys the next spring, and he was already in the woods in early May 1845. By August the crews had run over twelve hundred miles of lines and had made many "exceedingly interesting and valuable discoveries."[46] While Burt laid out township lines across the base of the Keweenaw peninsula, Houghton's efforts were concentrated farther up the point, subdividing the townships and fractions to permit accurate placement of the boundaries of the mining efforts that were ever increasing in number.

The summer of 1845 also saw the copper fever reach its peak. It seemed as if everyone was going copper hunting: "There are constant arrivals at Copper Harbor, and among them many distinguished men—ex-Governors, ex-Senators, ex-Secretaries, together with *Millionaires*—as well as geologists, surveyors, etc., etc., are all making a grand dash at the newly discovered copper region."[47] Simultaneously, Houghton achieved the pinnacle of his personal fame. The previous year he had been eulogized in an article published in several Michigan papers. It noted that "he has conferred great and lasting benefits upon Michigan by his professional skill and industry. No one man has done so much for any one State, in the way of scientific researches, as Dr. Houghton has for Michigan."[48] He was especially esteemed wherever he went in the copper country itself:

> Whenever he landed at one of the tent settlements—Eagle River, Eagle Harbor, Copper Harbor, or Ontonagon—he was immediately hailed and surrounded by the characteristically noisy miners. His small, boyish figure was easily recognizable among the burly frontier adventurers. He loved a good story, and so did they. And he was not too genteel for them. He dressed carelessly, talked fast, and was used to taking chances and living out-of-doors. And they respected him for knowing their business —copper-hunting on Lake Superior—better than anyone else.[49]

The Fateful Journey

The future for Houghton looked bright indeed. As summer yielded to autumn, he prepared to wind up the field season. On the morning of October 13, 1845, he embarked in a small boat from Eagle Harbor with four men, including his two favorite voyageurs, Peter McFarland and John Baptiste Boudrie.[50] They were carrying supplies for a survey party working along the coast some miles to the west. After stops and delays, they started back for Eagle River after sunset. The weather changed and it began to snow, but Houghton encouraged them to continue. The wind and the seas increased in violence and the boat was swamped. McFarland and Boudrie alone were able to reach the shore. Houghton, just turned thirty-six years old, disappeared beneath the icy waves along with two companions. His body, found only the next spring some six miles from the site of the drowning, was returned to Detroit and buried on May 15, 1846. His wife, Harriet, and two daughters were left without husband and father. One mourner spoke for them all: "The ways of Providence are truly mysterious!"[51]

Undoubtedly so. It is also true that after his tragic death, and perhaps even largely because of it, there began to grow up around Houghton's memory a worshipful haze that soon began to obscure the real man and his actual geological accomplishments. This has had far-reaching effects on the generally accepted picture of his significance in the development of American geology. Without doubt, this effect is most clear in the writings of Alvah Bradish, Houghton's friend, brother-in-law, portrait painter, and author of an influential account of his life.[52] Written in 1889, over forty years after Houghton's death, Bradish's *Memoir of Douglass Houghton* has played a major role in molding our present view of the geologist. Invariably, however, Bradish's concern was to place Houghton in as favorable a light as possible, both personally and as a geological pioneer. Yet, such admiration may not always do justice to its subject, and it may even be a barrier in trying to fairly assess the man on his own terms. A look at Bradish's analysis of certain events of Houghton's life provides an illuminating insight into the degree of confidence that can be placed in many of his interpretations.

After Houghton's death, it was inevitable that the question would arise whether it could have been avoided, whether carelessness had played a part in the disaster. Bradish strenuously objected to that implication, and defended Houghton by relating an event from the past that, he insisted, illustrated both the geologist's courage in facing danger and his resolute, deliberate nature: "He was never reckless at any time."[53] While Houghton's party was making its way westward along the south shore of Lake Superior at the beginning of the 1840 field season, the event took place which, apart from Bradish's claim, would merit little attention.

It was "related to the author [Bradish] by one of the men present." As Bradish tells it, the party was making its way along the coast, approaching the Pictured Rocks. A severe storm arose, which threatened to sweep them to disaster against the sheer, rocky cliffs. Houghton immediately recognized their danger, but he also knew that at rare intervals along the cliffs narrow openings existed that might afford shelter. "While the wind whistled and moaned, and the waves broke in thunder, leaping high up the cliff, the geologist stood firm at the helm . . . his eagle eye taking in the long range of sandstone cliffs, watching with intense eagerness," searching for that small opening in "the rocky walls that threatened any moment to receive them, to crush and grind them to pieces." At the last moment, "amid the lurid flashes of lightning" that illuminated "the pallid faces of those faithful assistants . . . that imperfect narrow opening appeared, and the frail bark with its precious freight was whirled in and shot up the slope, safe on the narrow gravel beach; and here these storm-beaten toilers of the sea remained until daylight."[54] Truly impressive, and done like a hero.

Now let us examine this episode more critically. On that survey of 1840 Houghton joined his party at Grand Marais on the afternoon of Thursday the eleventh of June, and they left there, continuing westward, on the twelfth. On the evening of Saturday the thirteenth, having passed the cliffs, they camped at Miners River at the west end of the Pictured Rocks. The evening of June 12, 1840, then, is the only possible date that will fit the chronology for the event referred to by Bradish.

Several members of the party kept regular journals of that 1840 expedition. Hubbard's entry for that date contains the simple statement that they "passed the Sables, & a storm coming on, encamped a few miles E. of the commencement of that range of lofty sandstone cliffs, which bears the appellation of Pictured Rocks."[55] Penny's notes elaborate further. He confirmed Hubbard's statement that they landed about three miles *east*, that is, *before* reaching the rocks and that the waves were high. "The process of landing was quite a novelty to a land-lubber like myself. As the bow of the boat touched the sand, the men sprang into the roaring surf, some holding the boat while others first took us on their backs to shore, and then took out the loading in the same manner. Then the boat itself had to be hauled up some rods on the beach." Finally, Houghton's own account of that evening states that "the wind blew a gale and compelled us to land upon a sandy beach with high *sand* cliffs in the rear [my emphasis]."[56]

And that is it. No waves thrashing against the sandstone cliffs, no narrow opening in the rock, in fact, no rocky cliffs at all; those were still several miles to the west. According to Bradish, the "storm-beaten toilers" spent the remainder of that wild night clinging to their narrow ledge of safety. In reality, once they had landed it was business as usual for the group. They warmed up by a "good fire," had some "refreshment," and afterward continued their investigations by walking several miles westward toward the sandstone cliffs. Houghton's own notes state that "after landing I walked along a sandy beach with sandy cliffs in the rear, westwardly for a distance of 1½ miles to that point where the elevated cliffs of sandstone forming the commencement of the Pictured rock range first make their appearance."[57] All this is not, in any way, an attempt to minimize the perils of their travels. Without doubt there were dangers aplenty in coasting the Great Lakes in small boats, but, objectively viewed, the episode seems to tells us much more about Bradish and the needs of his time for heroes than it does about the person of Houghton.

Bradish continued this approach in his treatment of the events surrounding the deaths in the survey group. It had been suggested by some, at that time, that carelessness by Houghton had contributed to the tragedy. "This idea would be absurdly unjust," wrote Bradish, who once again was most anxious to defend Houghton against anything that might possibly be construed as a blemish on his character. "He had the entire confidence of his boatmen and associates, who were bound to him like children to a parent . . . they were enthralled by the example of purpose, decision and energy, by confiding counsel and, in fine, by the unfailing resources of genius."[58]

If the accounts of those present at Houghton's death are examined, however, it is apparent that once again Bradish has obscured the real person of his subject behind a reverential facade. McFarland and Boudrie, the two survivors, were Houghton's favorite voyageurs. McFarland had traveled with him extensively and had wide experience on the lake, knowing its moods well. Before setting out on that last fatal leg of the journey, he told Houghton that he thought it was going to blow, but Houghton insisted on proceeding. Once under way and in heavy seas, McFarland asked Houghton to put in to shore, for they were then at a sandy stretch of beach where landing would have been possible, but again the answer was to continue. Boudrie then addressed McFarland in French, telling him that they must go ashore, but, after asking McFarland to translate Boudrie's remarks, Houghton once more insisted on pressing on.

After rolling around in the heavy seas and making little progress, McFarland gave Houghton a life preserver and advised him to put it on, but the geologist set it on the seat next to him. Finally, after nearly being swamped, Houghton decided it was time to try a landing, but McFarland told him it was too late, because the coast was now all rock. Houghton, in the stern steering as usual, immediately put about and the boat was swamped and capsized. On surfacing, McFarland grabbed Houghton by the collar and told him to take off his gloves and hold on to the keel, and "the advice was followed." Another large sea struck and they were separated. Of the five, only McFarland and Boudrie reached the rocky coast safely. Instead of illustrating a picture of children looking to a parent for guidance, it would appear that the final voyage was, in fact, initiated and continued by Houghton against the specific and repeatedly expressed better judgement of his voyageurs.[59]

One final, little noted incident may also be illuminating here. In the summer of 1845 James Hall, the great New York geologist, was sent to Isle Royale by a company interested in prospecting for copper. Delaying his departure from the island until late in the fall, he faced near mutiny from his crew who feared being isolated on the island for the winter. Hall finally set out for Keweenaw Point in a small sailboat loaded with twelve hundred pounds of specimens. After a harrowing, stormy crossing that required all of Hall's navigational skills born of his years of experience on the East Coast, he brought them to a dramatic but safe landfall.

Interestingly, that landfall was at Eagle River, the date was early in October 1845, and watching their arrival from the shore was Houghton, who could scarcely believe that such a trip would be possible under such adverse conditions and in so small a boat. Ten days later Houghton was dead. Hall, even after the passage of half a century, was still wondering if he might have been indirectly responsible for Houghton's death, in that his own dramatic crossing from Isle Royale may have encouraged Houghton to take the risk that ended up costing him his life.[60]

Whatever the exact circumstances, Houghton was gone and the entire state mourned. Michigan had lost not only its state geologist but one of its best loved citizens, and the first geological survey of Michigan came to an abrupt and ignominious end. Not the least of the consequences of the dismal event was Michigan's loss of control over the geological investigations of its own mineral resources in the Upper Peninsula. Despite the strenuous and active attempts by many in Michigan to prevent it, further investigations of the geology of the copper district passed into other hands, and came to be carried out largely by persons from beyond the borders of the state.

6

HOUGHTON—THE MISUNDERSTOOD PIONEER

The final report. In that single phrase is captured all of the initial bright promise and high expectations and also the subsequent bitter frustrations and ultimate failure of the work of Houghton and the first geological survey of Michigan. The final report was never written. The implications were many, but one of the most far-reaching is perhaps best expressed as a paradox: after Houghton's death he achieved an unrivaled reputation as *the* prophet who foresaw the value of Lake Superior native copper for profitable mining, and yet he left behind no easily accessible summary of his views either on Michigan geology in general or on the copper district in particular. The result has been that for a century and a half Houghton has been a misunderstood geological pioneer, and a significant part of the reputation he has acquired is not supported by a careful examination of the historical data.

His death sent waves of shock through Michigan and the rest of the country. Among those familiar with his geological efforts the sense of loss was particularly acute. A geologist wrote in Houghton's obituary in the *American Journal of Science* of "that melancholy event, which has deprived science of one of her most untiring votaries, the state of her geologist, and his friends of the most amiable and true-hearted of men."[1] In Washington, a House report mourned that "the sudden and untimely death of that learned and indefatigable gentleman has robbed science of one of its brightest ornaments, and the country of one of her most enterprising and useful citizens."[2] To Alexander Winchell, the second state geologist of Michigan, Houghton's "personal daring gave opportunity for the elements with which he had so often toyed, to blot his name, too early, from the list of America's most distinguished devotees of science.[3] Expressions of dismay over Houghton's death have spanned the years. Lewis Cass perhaps spoke for many of that time when he said that Michigan would be better off ten million dollars in debt rather than having lost Houghton,[4] and a recent judgment has called his demise "probably the greatest loss American geology had suffered in decades."[5]

There is little doubt that the feeling of loss was made particularly keen not only by Houghton's young age but by the distinct sense of unfinished business and unrealized potential that centered on the final report. Ever since the initiation of the state survey in 1837 that report had been the ultimate goal toward which everyone had been looking. It was to have been, above all else, a *practical* document, of direct use by the citizens of the state. All else to this point was only prologue. The annual reports, even the copper report itself, were only steps toward that goal, as Houghton himself had repeatedly made clear from the beginning. Throughout his years of work he was explicit about the importance of the final report, and his writings abounded in references to it.[6]

This culminating document was also widely anticipated by others, both within the state and across the country. While Houghton was still alive, after noting the work on the copper district that he had included in the annual reports, a geologist wrote in the *American Journal of Science* that "a full and detailed description of the minerals and mines of that country, will be published in his final report, which will appear on the completion of his extensive and arduous surveys."[7] Geological surveys were being completed in other states, and Houghton's final report was particularly anticipated for its analysis of the copper region and an expected realistic assessment of its potential for mining. By late spring of 1845 a Detroit newspaper, relying on what certainly must have been very "inside" information, confidently asserted that "the state work on geology will most probably be published this fall, and will embrace topography, botany, geology, mineralogy, and ornithology—two volumes octavo, illustrated with sections."[8]

Now, with his death, none of this was to be. The double loss of Houghton and the fruits of his labors was widely felt. J.

W. Foster and J. D. Whitney, who five years later produced the definitive report on the copper district, wrote in that report that "had he lived to complete this great work, he would have erected an enduring monument to perpetuate his name. He died in the discharge of his duty, prematurely for the cause of science, prematurely for his own fame."[9] It became a continuing theme that death had cheated both the world of the benefits of Houghton's final insights, and Houghton of the chance to firmly establish himself as among the greatest of American geologists.[10]

In the light of this anticipation Houghton's delays seem all the more curious. Yet, the press for the appearance of the final report does not seem to have been of any great concern to him, despite his pronouncements about its ultimate significance. In a letter to his father in 1841 he revealed that although the materials for the final report had been accumulating, even then, for five years, it would "not probably appear until two or three years from this time."[11] Whatever the exact reasons, at the time of Houghton's death there did not exist any document that could, even in a very preliminary sense, be called a final report. He had made little progress toward producing it, and notwithstanding several attempts to compile one after his death, no such report ever appeared.

How, then, can a fair evaluation of Houghton's geological accomplishments be made? In subsequent years the belief has occasionally been expressed that some of the results of Houghton's investigations were incorporated into the reports of later workers, but without full acknowledgment of him as their source. Therefore, it is said, he has not received the credit that is rightfully his for views that he held or for conclusions that he had drawn. Charles T. Jackson, J. W. Foster, and J. D. Whitney were all geologists who worked in the district in the years immediately after his death and whose reports will be examined more closely below. Such a claim has been explicitly made about the results of the efforts of these men: "Some of the assistants of Dr. Houghton were employed by Jackson, Foster and Whitney and Houghton's results were made use of in other ways to such an extent that a large part of the credit rightfully belongs to him although no proper acknowledgment of it is made by these geologists."[12]

It is clear, then, that the contributions of Houghton to our geological understanding of Michigan in general, and of the copper district in particular, are by no means obvious or easily determined. His matured and scientifically detailed views on these subjects at the end of his shortened life do not exist in any readily identifiable form. However, the annual reports, while explicitly not intended to be definitive or to include all the supporting details, do provide at least a partial account of his work in the copper region.[13] Perhaps even more significant, however, is that beginning at the very time of his death and continuing to the present, there has been a clear and explicit tradition about Houghton's geological accomplishments. This, in turn, has produced a generally accepted consensus on the content and significance of his views on the nature of the district and the copper found there. This tradition, which has come to dominate virtually all subsequent interpretations of his efforts, deserves closer examination, and will be considered in some detail here.

The Houghton Tradition

What I will be calling "the Houghton tradition" appears in what seems to be its earliest and most explicit form in the writings of John R. St. John. He was a well educated amateur who traveled extensively on Lake Superior during the summer of 1845, noting carefully and in great detail all that he saw and heard. The following year he published *A True Description of the Lake Superior Country*, which was perhaps the best of several such "guidebooks" being written for those who might be interested in seeking their fortune in the new "El dorado of the north."[14] His stated purpose was to make information about the district available to visitors to the region and to potential investors in the mining companies that were being organized, and in that work he included an account of how the uniqueness of the native copper had come to be generally acknowledged. He further made it clear that he was simply repeating, for the benefit of the uninitiated, a generally accepted story:

> If there is one fact which characterizes the bounty of nature to ours over the mineral of *all* other countries, it is that *fact and peculiarity* of our Lake Superior native copper, that it is in *no instance contaminated with alloys of other metals*. The assertion of which fact, when made by Dr. Houghton, was treated as a burlesque by scientific men at home and abroad, who called it "backwoods mineralogy."

His representations as to the great abundance of copper indicated by "surface appearances," were treated as "new country stories," and Dr. Houghton, smarting under this ridicule, pursued his researches for ten successive years before his reports elicited any public attention. He has gone down to his grave . . . leaving one point only fully established—that the accepted systems of geology and mineralogy are in many particulars inapplicable to the scene of his labors [his emphasis].[15]

Here we have, less than one year after his death, virtually all of the basic features of the "Houghton tradition": his clear recognition of the unique nature of the district and the presence of the native copper in great abundance; his unwillingness to accept outmoded theories based on mining experiences in Europe; the ridicule by his scientific contemporaries "at home and abroad"; the lonely labor of many years to prove himself right; and finally, at his death, the general recognition of the inadequacies of "accepted systems" and the validity of his conclusions about the native copper.

These same themes were soon reinforced by others. In 1848 Bela Hubbard, from the beginning Houghton's valued right-hand man and business associate, confirmed the heart of the story in a brief memoir of Houghton in the *American Journal of Science*. Hubbard noted that

the novelty of the geological positions advanced, and their apparent unconformity, in many particulars, with the state of facts existing in other well known mining districts, created only incredulity in those who had received their teaching in other schools of science. The Doctor was styled, in derision, a "backwoods geologist." But the progress of discovery since, though collected from numerous and independent sources, and opened to the scrutiny of many of the best scientific observers of the age, has confirmed the positions assumed by him; while so far from discrediting a single fact advanced, but little has been added even to those already made known.[16]

A similar picture was painted by an article published in a Flint, Michigan, newspaper in 1874, written twenty-nine years after Houghton's death:

When his report was first published his statements were attacked by the celebrated Prof. Silliman in his journal, and also by the leading geological savants of Europe. It required but two years working of the mines to satisfy the world that the "field geologist" of Michigan was right, and that some of the theories of parlor geology, however much they might be respected for their learning, must yield to the crucible test of actual demonstration.[17]

Further confirmation came from Alvah Bradish in 1889, whose general attitude toward Houghton has already been examined in the previous chapter. He wrote that Houghton's geological views had been widely condemned by his fellow scientists, saying that the "solons of the east" called him a "backwoods geologist" or "the boy who had a good deal yet to learn." Nevertheless, said Bradish, not a single fact stated or position taken by Houghton had subsequently been called into question, and the results of more recent investigations had "confirmed, with singular minuteness, all the theories propounded and every prediction he made."[18]

Frequent reference was also made to Houghton's "open mind," which contrasted so greatly with the closed-mindedness of his contemporaries. Bradish wrote that "he was not afraid of new paths if they would lead him to truth. Houghton was willing to see and accept facts with his own eyes and to apply them with honest convictions."[19] Another account stated that "he loved better to study the operations of nature in her own workshop than to read the theories of any man in his parlor. His mind could be satisfied only when he had put everything into the crucible and tested it for himself."[20] Indeed, Houghton had explicitly declared his intention not to be guided by past thinking, and "to be sure and not deceive myself, and to draw no conclusions but such as are strictly based on observation."[21]

In what is probably the most complete scholarly treatment of the Michigan copper rush period, this story remains unchanged. The author incorporated the "backwoods mineralogist" and "new country story" phrases that have become a standard part of Houghton historiography, and noted the special appeal that Houghton's views had for the common man:

> The claim that the masses were unadulterated "native" copper was received derisively by men of science, both at home and abroad, as a "new country story." Douglass Houghton acquired a reputation as a "backwoods mineralogist." One of [his] mental traits was a conscious attempt to keep an open mind. His report of 1841 upset traditional generalizations that had been derived from better known mining regions, particularly those of Europe. Whereas Houghton's fellow geologists received with incredulity his claim that on Lake Superior's shores unalloyed copper was to

be found not in rare instances, but commonly, the layman accepted this assertion at its face value, greeting it with "unalloyed" enthusiasm.[22]

And this interpretation continues today. The writer of a recent review of Houghton's life and work, after lauding him for his conclusion that native copper would be the basis for a successful mining industry and for his insights into the nature of the native copper lodes, said that Houghton was clearly "a creative and original thinker unfettered by the currently accepted explanations of the time."[23]

It would appear, then, that beginning with the testimony of Houghton's closest acquaintances and extending to the most recent analyses of his work, virtually unanimous consent has been reached on two themes. First, Houghton clearly recognized the abundance and potential of the unique native copper of the Keweenaw, despite the pointed opposition of the professionals of his day. Second, this recognition was in large part a result of his personal characteristics, which included an openness of mind and a freedom from inhibiting preconceptions, characteristics which contrasted with those of many of his professional contemporaries.

This story is certainly coherent, and it has appealing features: an engaging hero, a clearly defined claim, opposition from the major figures of a well-established scientific community, regional conflicts between the expanding west and the established east, a textbook confrontation of true empirical science versus outmoded authoritarianism and theory, years of lonely effort, and then a tragic death followed by final posthumous vindication. It sounds impressive, and much of Houghton's current reputation has been based on this view of his work. One further point might also be noted. If it is accurate, this picture raises a question about the analysis offered in previous chapters of this study. There I recognized two distinct groups between 1800 and 1850, the "amateurs," who were largely laymen who did recognize the significance of the native copper, and the "professionals," who largely rejected the possibility of its importance. More exactly, if this generally accepted interpretation of Houghton's work is correct, it would make him the first geologist to truly unify the two traditions, by the strength of his commitment to his observational data and despite his earlier geological education and the opposition of his colleagues.

A basic question nevertheless remains. As attractive as this heroic story is, does it fairly and accurately represent Houghton's role in the events that led to the recognition of the importance of Lake Superior's native copper?

Questioning the Tradition

Despite its coherence, persistent and nagging questions remain that make one wonder about the weight that should be given to this tradition. There seems to be no clear agreement among the purveyors of these views about the exact details of the story. St. John is obviously careless with dates in several places, and there is simply no way that his claim that Houghton conducted "research for ten successive years" can be fitted into any realistic chronology of Houghton's shortened life.[24] At the very least, St. John would seem to imply that the ridicule of Houghton came early in his career, presumably after his first report of 1832, to allow time for the ten years of work. Bradish, on the other hand, strongly implies that the "incredulity" of his contemporaries was a result of his copper report of 1840, published only five years before his death. Another curiosity is that the individual critics involved are often not explicitly named in many of these accounts. Instead, they are referred to rather vaguely as "scientific men, at home and abroad" or "the solons of the east" or "the leading geologists of the East that . . . laughed him to scorn."[25]

The exact nature of the conclusions that led to these alleged criticisms of Houghton is also not always clearly stated. To St. John, writing less than a year after Houghton's death, it was his views on the abundance and purity of the native copper that raised the ire of those unfamiliar with the region. To Bradish, it was the "novelty of the geological positions," which did not agree with other mining districts of the world. To Martin, writing over a century later, it was his "theory of the formation, structure, and position of the copper bearing rocks." Unfortunately, the story in all these accounts is almost completely undocumented. The "backwoods mineralogy," "boy geologist," and "new country story" phrases that sound so catchy are presented as if they are quotes, but no sources are given.

Bradish, writing forty-four years after Houghton's death, did fill in some further specific, but again undocumented, details. He claimed that the opposition began as a result of

Houghton's presentation of the results of his work on the copper district to a meeting of the "American Geological Society," in Albany, New York, by which he undoubtedly meant the Association of American Geologists and Naturalists, which would have been in 1843.[26] Martin, however, writing much more recently, states that it was a result of the meeting at Philadelphia, which would have been in 1841.[27] Bradish also did name some of the persons involved: "It should be remembered that the association was composed of such men as Prof. Hall, the Rogers brothers, of Pennsylvania, Dana, Silliman, Torr[e]y and many others, many of whom were among the most distinguished of living geologists. They would naturally be somewhat incredulous of these new and dubious theories of the 'backwoods geologist of Michigan.'"[28]

It is difficult to evaluate accurately this rather vague perspective on the reception of Houghton's work, but some of its implications do not ring true. The contrast of Houghton, the "backwoods boy geologist" with "the solons of the east," for example, seems overdrawn. It could just as easily have been argued that Houghton himself was one of those Easterners. After all, he had gone west himself from an eastern state, and he had graduated, *before* the great geologist James Hall did, from the most respected geological school in the country. His connections in the East had been strong ever since his college days. Being part of the expansion of the nation westward was no basis for reprobation; it was part of the high calling of the day. Trips to and news from the western lands were very much a part of the national life of that time. He had maintained very active scientific contacts in the East, among them with Torrey, one of the persons mentioned by Bradish, but whose high opinion of and respect for Houghton have been noted previously several times.

Besides his letters, Houghton's main interactions with eastern scientists would have been through the annual meetings of the American Association of Geologists and Naturalists. Yet Houghton was not only an active member of that organization but a full participant in the meetings in every sense. Accounts of those meetings were published regularly in the *American Journal of Science* from 1840 to 1845, even to the point of including details of questions and answers by the participants, among whom were many of the most prominent geologists of the East. It is difficult to imagine anything approximating the attitude described above when reading of the extent of Houghton's participation in those meetings, his service on several committees, and his election as treasurer of the organization.[29]

Specifically cited by Bradish as opposing Houghton was Benjamin Silliman, founder and editor of the *American Journal of Science*: "Perhaps even the accomplished and learned editor of 'Silliman's Journal of Science' may be suspected at this time of a want of that catholic spirit of liberality that should always characterize a scientist."[30] And yet Houghton's activities as state geologist of Michigan were reported early and favorably in Silliman's journal. In 1838, for example, the journal noted of Michigan that "this young State has set a laudable example, in ordering a geological survey, under Dr. Douglass Houghton, which he has carried on with particular zeal." Of the difficulty Houghton was having in obtaining funds, it was opined: "This ought not so to be . . . especially where they are so well deserved."[31] On the first anniversary of the Association of American Geologists, at the same meeting at which Houghton made his major presentation of his work on the copper region, Edward Hitchcock delivered a significant address on the state of American geology and its development, which was also printed in the journal. In it, Houghton was mentioned several times in a completely uncontroversial manner, as one whose opinions on geological questions were a normal part of the growth of the science of the day.[32]

The charge directed against Silliman appears to be the only instance in which one person is singled out for identification as an opponent of the ideas of Houghton. It would certainly seem that something must have been written in the *American Journal of Science* to inspire such a charge. Yet, an examination of the journal from 1840 to 1845 did not reveal anything that might be construed as such a criticism.[33] Silliman, as a leader of considerable influence in the geological community and one particularly interested in its development of a professional image, went out of his way to try to ensure that any differences of opinion would not lead to personal controversies.

In 1842 Silliman made a major address to the association, in which he emphasized that American geologists should always act toward one another with "justice, honor, fidelity and delicacy," addressing one another "with the strictest rules of geological courtesy," so that their interactions might become "both focal and radiant points of intellectual light and moral influence, to the honor of our country and the common benefit of mankind."³⁴ This is not to suggest that controversies were absent from the meetings of the association, and at some there were distinctly heated exchanges. Silliman, however, was very conscious of the public image he felt the group should present, and he hardly seems a likely candidate to have openly demeaned the work of a colleague.

There are several additional reasons for questioning the accuracy of the "Houghton tradition." The native copper had received extensive comment long before Houghton's time. Schoolcraft's earlier views on the abundance of native copper had been widely disseminated. About 1820, pieces of the native copper mass that had been analyzed in Utrecht had been distributed in Europe, and had also earned comments in several European texts.³⁵ Schoolcraft's report had been translated and published in Germany in 1822.³⁶ In 1826 Schoolcraft had received an offer from Austria to exchange a large mass of native copper from his collection for minerals from the Imperial Cabinet of Vienna.³⁷ Parker Cleaveland, in the second edition of his important pioneering mineralogy text of 1822, included a summary of Schoolcraft's observations on the Ontonagon Boulder under the entry for native copper, and he noted that a comparable native mass was known from Brazil. He also mentioned the smaller masses so frequently found along the coasts of the Keweenaw, and the analysis of the copper at Utrecht.³⁸

As for the opposition from the "scientists at home," we have seen that here in the United States the district had already attracted great attention at the highest levels of government as far back as the very early 1800s. Its potential had been frequently noted long before Houghton's first visit in 1831. The *American Journal of Science* had noted in 1834 that "the numerous and important facts mentioned by Mr. Schoolcraft [in 1821], render it certain, that native copper is frequently found in that region, and lead to a strong presumption of the existence of valuable mines of copper."³⁹

There were two influential mineralogy texts that appeared in America between Houghton's first visit to the copper country in 1831 and his death in 1845, and which therefore might possibly have reflected a published opinion on the question of native copper in the Keweenaw. They were those of Charles Shepard, in two parts in 1832 and 1835, and that of James Dana in 1837. About native copper, Shepard explained that it was found in detached masses in several places in North America, including Michigan and the Northwest Territory. He noted that the mass on the "Ontanawgaw" River was still lying there in the riverbed and that there was a large fragment from this mass in the collection at Yale.[40] Dana, one of the "incredulous" according to Bradish, had noted in his first edition of what was to become the "bible" of mineralogy in the United States, that native copper had long been known from many parts of the globe. He also spoke of the "magnificent mass" on Lake Superior and included Schoolcraft's estimate of its weight, noting further that "smaller masses are quite common in the same region."[41] In neither account was there any indication of a questioning of the nature of the district, or any indication that the native copper was not in fact present exactly as it had been described.

The general impression gained from what has been examined here, then, does not seem to inspire confidence in the view sketched by St. John, Hubbard, and Bradish, despite its continued acceptance down through the years. However, demonstrating a negative, that there was no such criticism of Houghton, would be a very difficult thing to do; virtually all published accounts touching on native copper between 1832 and 1845 would have to be examined, and such criticism, even if made, might not have been committed to print.

It just may be, however, that we are trying to do the impossible because we are asking what may not be answerable. Perhaps a more fruitful approach would be to reverse the traditional perspective and ask instead two alternative questions. First, is it true that all Houghton's fundamental conclusions about the geology of the district have been confirmed and supported by later work, and second, is it true that Houghton believed that native copper was present in economically significant amounts in the Keweenaw? If we are willing to consider the possibility that the Houghton tradition may be incorrect, the entire issue then becomes much more accessible. The answers can now be sought not by scrutinizing volumes of obscure publications, searching for the identities of Houghton's alleged critics and their criticisms, but by simply looking more closely into such writings of Houghton himself that we do have.

Testing the Tradition

It is undoubtedly true that the studies of Houghton were basic to our present understanding of the geology of the copper district. Without doubt they provided a good foundation on which later workers could build. At the same time it is also true that later geologists drew conclusions that went beyond, and in some cases were distinctly contrary to, those reached by Houghton, and some investigators have not been reluctant to note such instances where appropriate.[42] However, the temptation to claim that all Houghton's conclusions have been confirmed, and to attribute to him geological views more advanced than those he held, has proven to be almost irresistible for many writers down through the years. As a result, the true nature of many of Houghton's geological contributions has been considerably obscured.

In 1982 the first issue of the new journal *Earth Sciences History* included an article on Houghton's life and work that, after a long delay, has once again brought Houghton to the attention of a national audience.[43] The conclusions expressed in this article were repeated in an even more recent account of Houghton that appeared in 1987 in *Geotimes*, another national magazine that featured a portrait of Houghton on its cover in memory of the founding of the first geological survey of Michigan 150 years earlier.[44] These articles provide examples of how Houghton's work in the copper district is being interpreted by some authors in the light of our present knowledge of the geological events that brought those rocks into existence.

In the first of these articles, Houghton's contributions to our understanding of the geology of the copper district is summarized in this way:

> [Houghton] determine[d] the exact location of the copper deposits, its various copper ores, and the nature of the different copper occurrences within the district. . . . Houghton determined that the major native copper deposits were associated with a Primary (Keweenawan) volcanic series consisting of about 400 distinct basaltic lava flows containing 20 to 30 interbedded conglomerates, all steeply dipping to the northwest. His examination showed that the native copper always occurred in any of three situations within the Keweenawan basaltic lava flow sequence: in fissures within the flows, in the amygdaloidal zone at the top of a lava flow, or in the interflow conglomerate intercalated between two lava flows. Furthermore, he noticed that the copper always occupied the topmost portion of the permeable zone just beneath an impermeable zone—usually the overlying basalt flow.[45]

The important question of Houghton's views about the copper will be dealt with more fully below, but virtually every claim made here for what Houghton supposedly discovered is seriously deficient or simply incorrect.[46] In particular, Houghton never concluded that the traprocks of the Keweenaw were "lava flows" or that they were part of any "volcanic series." Instead, Houghton consistently called them "dykes"; that is, he believed that they did not form by lava pouring out onto the surface of the earth but by molten traprock rising upward from below and forcing its way between beds of preexistent strata in the overlying rocks. The difference is not trivial; in Houghton's view, the conglomerates and the overlying younger rock layers were already there even *before* the traprocks were formed, a conclusion the opposite of that reached by modern geologists.

Examples such as this are not being cited to in any way demean the efforts of Houghton or his assistants. The work they did under difficult conditions was truly remarkable, and we should never fail to freely acknowledge the contributions of those who have gone before us. At the same time, a clearer understanding of the past is not well served if misplaced admiration leads us to attribute to our predecessors things they did not in fact know or interpret correctly by modern standards. Much of what may seem clear and unambiguous to us today was not always so in the past.

The major question of what Houghton discovered, however, does not lie in what might be considered the minor details of geological interpretation. Our primary concern must be the question of the value of the native copper and the degree to which Houghton did or did not recognize it. It is here, however, that a closer look at Houghton's writings will reveal what are the strongest reasons for questioning traditional interpretations of his work. We will therefore go back and examine more critically some of Houghton's statements about the copper, but in doing so it is important that the various uses of the word *ore* be kept clearly in mind.

In the strict, technical sense the word *ore* by itself does not imply anything at all about the chemical characteristics of a mineral. The word may be used with equal validity for copper in either the native state or for minerals in which the metal is chemically combined with other elements such as sulfur or oxygen. In practice, however, during the time that we are considering here, the term *ore* was frequently used in a narrower sense, referring only to copper in its chemically combined state, as, for example, in a sulfide, oxide, or carbonate, but not to the metal in its pure, native condition (a usage that continues to some degree even today). Further complicating the picture is an inconsistency in even this usage, so that in any consideration of the nature of the copper a sensitivity to exactly what meaning is being given to the word *ore* and the context of the comments being made is important. Some extended quotations will therefore be examined.

Perhaps the earliest significant statement by Houghton on the copper deposits is the letter to territorial governor Lewis Cass that was the official report of the Schoolcraft expedition of 1831. In it he clearly recognized the traprock as the original source of the copper masses, such as the Ontonagon Boulder, that had for so long been found lying scattered over the land. In reading the entire letter it is clear that he did not, at that time, foresee the native copper, even if it could have been located within the rock itself, as a likely possibility for mining. He spoke instead of the ores that are "ordinary in appearance." Clearly, at that point, Houghton was thinking of copper compounds; "examinations for the ores of copper could not be made in that country with hopes of success, except in the trap-rock itself," but "what quantity of ore the trap-rock of Keweena [sic] Point may be capable of producing, can only be determined by minute and laborious examination."[47]

This occasion was his introduction to the district, and certainly cannot be taken as his matured views on the matter. Subsequently, as state geologist, he became much more familiar with the region. Nine years later he spent most of the summer of 1840 in the field, collecting the data he used in preparing his "copper report" of 1841. All agree that this report was the single most important factor in precipitating the mining rush. Yet, even then, his views do not seem to have changed. While examining the famous "green rock" vein at Copper Harbor on that trip, Charles Penny gave a fairly detailed picture of their investigations in his journal. After they had blasted at one vein, Penny noted that it contained "pure Malachite, green oxyd, and copper black," which are all compounds of copper, and then added the comment, "One *favorable* symptom is, that no native copper has been found in or near the vein [my emphasis]." This is a clear reflection of the traditional view, according to which the pure metal was not the form of copper to be sought, and this almost certainly reflects Houghton's thinking on the issue, for Penny explicitly stated that he was depending on "the Doctor" for the correct interpretation of what they were seeing.[48]

Later, in the middle of that expedition, Houghton sent a letter from La Pointe on Lake Superior to Michigan's Governor Woodbridge in which he said that the ores he was finding "are in a condition that they would scarcely be recognized by a superficial examination, and added the curious comment, "The difficulty is that our people are looking for copper in its native state."[49] Then, back home in December 1840 after returning from that most important field season, he wrote Woodbridge another letter, noted in another context in the previous chapter. His comments on the relationship between the native copper and the ores are revealing:

> But little, in fact almost nothing has, heretofore been known of what in reality constitutes the true mineral region of the upper peninsula. Loose pieces of native copper were occasionally picked up in the vicinity of Lake Superior and this led to the general belief that the metal existed in quantities in that region, but nothing definite was known upon the subject, for the reason that nearly the whole of the metalliferous rocks are situated in the interior, and in those rugged districts which heretofore have scarcely been visited by whites, and *so little resemblance do most of the ores which occur, in the veins alluded to, bear to the native metals, that the Indians would scarcely detect them* [my emphasis].[50]

That same month he wrote another letter, this time to Michigan congressman Augustus Porter, which contained the provocative and often-cited comment about blasting out two tons of ore along with a large piece of native copper weighing forty pounds. The context, however, demonstrates a different perspective:

> I brought from Lake Superior on my return to Detroit this fall from four to five tons of copper ores, and am now busily engaged in making an analysis of them. Thus far they have proved equal to any ores I have ever seen, and their value for reduction cannot be doubted. The average per cent of metal is considerably above that of Cornwall. While speaking of the ores I am reminded of the beautiful specimens of native copper that came out with the ores in opening some of these veins. They are not very abundant, but some of them are very fine. In opening a vein with a single blast, I threw out nearly two tons of ore, and with this were many masses of native copper, from the most minute specks to about forty pounds in weight.[51]

Again, this is revealing, even beyond the industry required to carry tons of material from the wilds of the Upper Peninsula back to Detroit. He clearly distinguishes between "ores" and "native," his comparison is to Cornwall where all the ores are copper compounds, his inference of value is for "reduction," which would be necessary only for copper combined with other elements, and finally the native copper, despite its spectacular appearance, is "not very abundant." It should be particularly noted here that Houghton cannot be using the word ore to refer to native copper, since he contrasts the "ore" that he found with other rock containing even the "most minute specks" of copper in the native state.

That next February the copper report appeared, which has ever since constituted his chief claim to fame. It continued this same approach, although this can be overlooked in casual reading. His point of comparison was always Cornwall, which is understandable in the light of its significance in world copper production at that time. It seems clear that Eaton's principle of analogy, which Houghton learned in his student days at Rensselaer, was influencing his thinking as he wrote about the copper minerals. Both compounds of copper and native, he said, "are distributed in bunches, strings, and comparatively narrow sub-veins, in a manner precisely analogous to that in which these ores are usually distributed, in similar rocks in other portions of the globe."[52] The difficulty was that in Cornwall and the other districts the only minerals that were present in amounts great enough to be ores were all copper compounds, precisely what they are not in Michigan.

He continued with his analysis: "As a general rule, those metals which are oxydable at ordinary temperatures, or which readily combine with sulfur, *rarely occur in a metallic state*, but are usually found in combination either with sulfur, oxygen, or acids [his emphasis]. . . . In the main, the resemblance between the character and contents of the copper veins of Cornwall and Michigan, so far as can be determined, is close."[53] The compounds of copper were clearly on his mind when he said "many of the richest ores are so far from having the appearance of the pure metal that they would be the last suspected to contain it in any form."[54] He recognized that the Cornwall and Michigan districts did differ in some respects. The main ore in Cornwall was the sulfide of copper; in Michigan, along with the native copper, there were mostly "oxydes and carbonates." He did recognize that native masses of all sizes were present, but of interest is his understanding of what this pure copper means.

After addressing those veins in which copper minerals were found, he turned his attention to those portions of the veins where the metal was concentrated in pure masses "from the merest speck to that of several pounds in weight." The concentration of so much native copper can be very misleading, he warned, calling the presence of the metal in excess of the other ores "a peculiar feature." He struggled valiantly trying to make the native copper he was seeing fit into his understanding of its minor significance in the more familiar, established mining districts of the world. It was, he said, "a condition which is similar to that observed in those veins of copper that have been extensively worked and found to be the most productive, on the continent of Europe and the Island of Great Britain."[55] Still, he had difficulty bringing himself to fully accept his own interpretation:

> The occurrence of this native copper in the veins, and the manner in which it is associated with the veinstones, in all respects corresponds with the ordinary association of the other forms of ores, in those veins that have been extensively worked in other portions of the globe; but I confess that the preponderance of native to the other forms of copper, was regarded as an unfavorable indication, at least until this had been found to be more or less universal with respect to all the veins.[56]

The latter part of this statement should be noted carefully. It can easily be misconstrued, and some writers have built upon it an interpretation of how Houghton came to understand the significance of the native copper. According to this interpretation, Houghton started out by accepting the traditional view that the native copper would not be present in amounts capable of being profitably mined, but then changed his thinking after finding that the great abundance of the native metal was "more or less universal with respect to all the veins." Therefore (it has been claimed), Houghton's thinking changed as a result of his observations in the district. He abandoned the ideas of the past and recognized, when he wrote the copper report, the true value of the native copper. He thereby (it would appear) stands clearly as the forerunner who pointed the way to the uniqueness of the district as a producer of native copper, even though this was not demonstrated until after his death.[57]

In reality, however, this is not the case. Houghton did have a change of mind, but it was much more modest than this, especially when it is viewed in the light of the entire report. Previously, he had considered the native copper to be a distinctly *negative* indicator for the possibility of mining the ores of copper compounds; that is, he regarded the native copper as counting against the possibility of ores of *any* type being present in amounts sufficient to mine. Now, because of its ubiquitous nature, he reluctantly conceded that native copper may not necessarily be viewed in so negative a light. There is no indication here, however, that he in any sense anticipated that the native copper would *itself* be "mineable."

As he approached the end of this most important section of his report, he commented briefly on the "fatigues and exposures" of the field season that affected his health and prevented him from completing his analyses of the ores he had brought back. Nevertheless, he said that "sufficient has been done to show satisfactorily that the copper ores are not only of superior quality, but also that their associations are such as to render them easily reduced [once again, "reduction" is required only for copper compounds]."[58] In giving his preliminary numbers for the percentage of copper in the ores, so as to facilitate his comparison of the percentages to those of Cornwall, he deliberately omitted the native copper from the totals. He also explained that his reason for not mentioning the Ontonagon Boulder earlier in his account, despite its spectacular nature and the role it had played in attracting attention to the region, was to avoid the possibility that "perhaps this isolated mass might be confounded with the products of the veins of the mineral district."[59]

Then, near the end of the report, he summarized both his view of the district and his conclusions about the likelihood of its commercial development. This is the most complete single statement Houghton ever made on the potential of the Keweenaw for mining:

> The value of a vein may be said to depend upon the abundance of the ore, and the ease with which it can be raised and smelted, rather than upon its purity or richness. Upon this point, with respect to our own mineral region, public opinion would perhaps be more in error than upon any other, and most certainly we could hardly look for a mineral district where the character of the ores was more liable to disseminate and keep alive such errors. The occurrence of masses of native metal, either transport-

ed or in place, is liable to excite, with those who have not reflected upon the subject, expectations which can never be realized, for while in truth, the former [isolated surface masses] show nothing but their own bare existence, the latter [native masses in the rock] may be, as is frequently the case, simply imbedded masses, perfectly separated from all other minerals, or they may be associated in a vein where every comparison would lead to unfavorable conclusions, as to the existence of copper, in any considerable quantities. I have frequently noticed very considerable masses of native copper, occupying the joints of compact greenstone, under such circumstances as I conceive might readily excite in many minds, high expectations, but a little reflection would satisfy the most careless observer of the uselessness of exploring these joints, under the expectation or hope of finding them a valuable repository of the metal. . . . While I am fully satisfied that the mineral district of our state will prove a source of eventual and steadily increasing wealth to our people, I cannot fail to have before me the fear that it may prove the ruin of hundreds of adventurers, who will visit it with expectations never to be realized.[60]

This extended quotation deserves careful examination. It is not an advanced, forward-looking statement that declares the region to be unique, differing fundamentally from all others in the world. Instead, this is a view that was very traditional in its essentials, very much in keeping with accepted theories, and in some ways even reactionary. Ironically, according to Houghton it was the very *presence* of native copper that was creating the errors in the public mind. One conclusion he drew is particularly interesting in the light of the widely accepted view today that Houghton was strictly empirical and observational in his outlook on nature, basing his conclusions only on what he himself could see for himself. He stated that "a little reflection" is the appropriate cure for the "careless observer"; that is, a little thought would cure the observer tempted to conclude that native copper *is* present in significant amounts! Houghton's repeatedly expressed and certainly prophetic warnings that in the coming rush many more fortunes would be lost than made, are often cited as an example of his far-reaching insight. Very seldom is it noted that these fears were, in his mind, motivated primarily by the belief that the public was misled in its estimation of the importance of the native copper. He struggled long and hard to fit this odd mineral into the conceptual boxes of his mind, but its meaning and importance continued to elude him.

It is further evident that these ideas, first expressed as a result of his first visit to the district in 1831 and confirmed by his detailed observations of 1840, remained the basis of Houghton's understanding of the district through the remaining years of his life. It is also clear that others outside the district adopted similar views on the greater merits of the ores of compounds rather than the native metal, at least in part, because of the weight of Houghton's reputation. As the acknowledged expert on the region, his views were solicited and respected both in his home state and throughout the country, and when he spoke out, the theme was unchanged.

A clear example of this is provided by notes taken at a meeting held in Detroit in 1843. The falls of the St. Marys River had long been a bottleneck to communication with the copper region. The building of a canal was inevitable, given the need for more direct access to Lake Superior and the pressures that were building because of the increasing attention given to the copper region. In April 1844 the "Committee on Roads and Canals" reported to the U.S. Senate its findings about the desirability of such a canal, and that report included notes taken at a widely advertised public meeting had been held in Detroit on November 23, 1843, to discuss reasons why such a canal should be built.

At that meeting, Houghton made extensive comments on the canal's desirability. A detailed account of those comments was published in a Detroit newspaper, and that account was then incorporated into the Senate report. He discussed his survey of the canal route and estimated that its cost "cannot at the utmost exceed two hundred thousand dollars."[61] He then commented extensively on the justification for such a canal and the reasons he believed its construction would be in the national interest. The copper region played a major role in his arguments, and of interest are his statements on the nature of the ores, as recorded by the reporter for the newspaper. Here, late in 1843, Houghton spoke before an audience of his fellow citizens. He expressed to them in plain terms the nature of the copper district and its ores as he, the state geologist and acknowledged expert, understood it:

> The ores chiefly consist of the carbonate, silicate, and oxides of copper . . . and he [Houghton] has as yet scarcely seen a trace of either the sulphuret or pyritous copper in this district. He spoke of this as a circumstance which had led to some error with those persons familiar with these last-mentioned ores of copper;

for, while looking for those which do not exist, they have passed by the dark and earthy-appearing mixed oxides of copper, supposing them to be inferior ores of manganese and iron. . . . He said he was well aware of the infatuation which existed over the whole world in relation to subjects of this kind, and that he felt it his duty to say that he hoped . . . no one would engage in the undertaking without "counting the cost," or without being fully informed upon the subject. . . . He further remarked, that the amount of native copper occasionally found in masses upon the surface, and at other times imbedded among the ores in the veins, had served to mislead the minds of many persons, who, having little knowledge upon the subject, have not stopped to reflect that copper in this form is of comparatively rare occurrence, and that in a practical point of view it is of no importance.[62]

A plainer or more straightforward analysis could scarcely be imagined, and its implications are clear. Well over twenty years earlier, on the Cass expedition of 1820, Schoolcraft had grappled with the same question of the meaning of the native copper. In doing so, even after having become aware of the views of other geologists, Schoolcraft had nevertheless still argued that the native copper would make a significant contribution to the success of mining efforts, even if other ores were ultimately the primary product. With Houghton there is no such allowance. He not only declared that the native copper, while spectacular and attention-getting, was not present in meaningful amounts, but also that it had misled and misdirected efforts that might have been more profitably directed toward the various other ores.

There is little doubt that these opinions of Houghton had a widespread influence. The possibilities of full-scale mining were increasingly being brought to the attention of Washington, and his reputation as *the* expert was universal. In April 1844 the Senate requested Secretary of War William Wilkins to pass along a report on the working of copper ores made by Capt. G. W. Hughes of the Topographical Engineers, because of "the great interest which has been recently manifested in the development of the mineral resources of the United States, especially in reference to our copper mines."[63] Hughes had twice visited Cornwall to collect "authentic information" on copper mining. In his report his discussion was entirely of ores and smelting, and he quoted Houghton to the effect that "the carbonates of copper are the predominating ores of that metal found on the southern shores of Lake Superior."

In February 1845 the Senate again requested any information recently received from General Cunningham, the mineral agent in the district. Again the report was all of "black oxide, grey copper, sulphuret, and blue and green carbonate."[64] The large specimens of native copper were mentioned, but there immediately followed the analyses from Houghton's report, which specifically omitted the native metal. The next month the Senate called yet again for more information, and got it in the form of the report of John Stockton, the new superintendent of the mineral lands on Superior.[65] Once more, the report ended with "ores" and "fluxes" and the fact that at Sault Ste. Marie "great facilities are offered for the erection of furnaces and the reduction of the ores. . . ."[66] As late as May 1846, over thirty-two pages of a congressional document were used for including the heart of the copper report of 1840.[67]

And yet, all this talk and concern about copper "ores" notwithstanding, well over a century of detailed examination since Houghton's death has shown that such ores are simply not present in significant quantities in any of the rock formations that he studied. What, then, could he possibly have been referring to? If his analytical results are reliable, it seems likely that Houghton had placed special and perhaps excessive emphasis on those samples he collected from the famed green rock vein at Copper Harbor. The landward extension of that vein does contain copper compounds, including dark copper oxides, and he and his party spent much time there collecting during the 1840 field season. The green rock was perhaps the most studied locality, after that of the Ontonagon Boulder, in the entire district, but it was also highly misleading. We now know that it was not at all representative or typical of the district as a whole, and the ores found there were present in only trivial amounts.

Houghton—An Alternative Perspective

There emerges from all this, then, a distinctly different picture from the one traditionally accepted for the relationship between Douglass Houghton and the copper deposits of the Keweenaw. Far from being the isolated visionary of the true nature of the copper, his thinking seems to have been very much in accord with conventional theories. The defenders of Houghton's role as the discoverer of the importance of the native copper are on the right track in their understanding of the conservative views of the professional tradition. They err, however, in not realizing that Houghton was fully a part of that tradition, in training, in background, and also in his views on native copper. Furthermore, these ideas seem to have been in place in his mind when he first visited the district in 1831, and the evidence examined here does not suggest these views changed in any meaningful way before his death in 1845.[68]

There is, then, a fundamental paradox at the heart of Houghton's efforts as he struggled to come to grips with the nature of the district. On the one hand, a lack of interest in useless theorizing and an explicit desire to be guided only by the facts were outstanding characteristics of Houghton's personality, noted by virtually all who knew him best. All seem to agree that he was open-minded, unfettered by outmoded theories and attitudes. Nevertheless, it would seem that this conventional picture is significantly flawed.

In Houghton's mind, there was one safe way to proceed in analyzing the new district: by using analogy, proceeding from the known to the unknown. This principle had been explicitly enunciated and applied to the search for minerals by Amos Eaton, Houghton's teacher, in the textbook he wrote while Houghton attended Rensselaer. It is undoubtedly a useful and even necessary methodological principle, and Houghton used it repeatedly in his comparisons of the Keweenaw and the Cornwall copper mining regions. However, in so doing he became deeply, if unknowingly, influenced by the conventional wisdom of the day, which said that native copper was not a mineral that could be profitably mined. At best, it would seem that he finally came to regard it as a not unfavorable indication of the ultimate success of mining other forms of ore. He evidently retained this opinion right up to his untimely death, which occurred, with great irony, just a few months before immense masses of native copper began pouring from the newly opened Cliff Mine. Without doubt Houghton knew the district intimately, better than anyone else of his time, but he was operating within an unacknowledged mental framework.

Bradish, almost in passing and undoubtedly without understanding its full implications, captured what may well be a very illuminating component of Houghton's personality when he wrote: "Still the Doctor was shy of new theories or mere speculations in science. This feeling might even reach to overcaution or dread. He was by nature cautious, and his hesitation to accept speculative views in science will be remembered as a marked trait in his whole career."[69] Houghton claimed to distrust the theories of the past, but he also feared or even "dreaded" the possibility that new theories might be called for. He chose instead to remain, so he thought, committed to strict observation alone, but no presupposition exhibits more subtle and unanticipated effects than that of which one is entirely unaware.

In a widely noted study of the nature of the scientific enterprise, W. I. B. Beveridge has written of the natural reluctance of the human mind to easily assimilate ideas that conflict with deeply held beliefs:

> In nearly all matters the human mind has a strong tendency to judge in the light of its own experience, knowledge and prejudices rather than on the evidence presented. Thus new ideas are judged in the light of prevailing beliefs. If the ideas are too revolutionary, that is to say, if they depart too far from reigning theories and cannot be fitted into the current body of knowledge, they will not be acceptable. When discoveries are made before their time they are almost certain to be ignored or meet with opposition which is too strong to be overcome.[70]

Here, in the present episode, this analysis can be shown to cut even deeper. Houghton was not faced with someone else's "new idea" or "new theory." It was nature itself that was "too revolutionary," that "could not be fitted into the current body of knowledge." If the abundance of native copper and the absence of copper compound ores can together be regarded as a "fact" (and this was emphatically demonstrated over the century that followed Houghton's death), it is clear that not only can it be difficult for a mind to assimilate new ideas, but even the very act of perceiving a "fact" can be a perilous enterprise. The observations that Houghton made so carefully and which he honestly felt were so objectively based, were able to take on meaning for him only to the degree that they could be integrated with his existing mindset. The educational background that led him into geology, with its emphasis on analogy as applied to the mineral districts of the world, had a far greater hold on him than he ever realized.

A last question remains. Why did Houghton become the central figure in this story that completely inverted what really happened, and why would it endure for well over a century? It may well be impossible, at this late date, to provide definitive answers to this question, but several possible contributing factors can be suggested, perhaps none of which if taken alone would have sufficed.[71]

We have already seen that the period during which the mining potential of the Keweenaw was being realized was a time of transition for science in America from an amateur to a more professional perspective. Many of those who contributed to the Houghton tradition were themselves amateurs. They were, to at least some degree, ambivalent about the emergence of science as a professional discipline, because it implied that the common man could no longer judge for himself the nature of what he was seeing. An undercurrent of latent hostility seems obvious in the use of such terms to describe the professionals as "savants of Europe," "solons of the East," and "big heads."

Houghton had never tried to conceal his opinions on the value of the native copper. However, the tragic aspects of his death, the lack of a final report, and the brevity of his earlier annual reports left, perhaps, some room for uncertainty about what his views were. Houghton had also, perhaps unwittingly, made himself available in other ways. Although he was one of the emerging professionals in geology, he had always been a favorite of the common prospectors and miners. He liked their company and moved easily among them and they strongly identified with him. His unpretentious nature and easygoing personality contrasted favorably with their mental image of what the other professionals were like: "the little, rough-looking Doctor carried more true knowledge in his cranium 'than all the big heads put together.'"[72] Houghton's enduring appeal to the common man is dramatically illustrated by a tribute given by a state senator from Michigan in 1875, thirty years after his death:

> We have here no enormous London, no rich and cultured people bowing in enthusiasm before the throne of intellect, science, genius, and heroism. . . . We are a rough, practical, money-making race. . . . But we have great, warm, working, western hearts, which the icy waters that were his winding-sheet cannot chill, and they shall be our Westminster Abbey, Douglass Houghton's mausoleum. We will fix our eyes on his noble life and death, and by striving with generous ardor to emulate them, erect to him the imperishable memorial which history ever grants the teacher, and, perhaps by God's grace, may follow him to heaven.[73]

If Houghton had not tragically lost his life at just that critical time, he undoubtedly would have willingly, and even delightedly, expressed his amazement at seeing the native copper that would soon come pouring out of the mines. In that case, it would have been clear to all that his views, like those of almost all the other professional geologists of that day, were being shown to be inadequate, and there would have been no "Houghton tradition." But Houghton's death did occur, and after his death it was easy for many, with slight mental adjustments, to transform the man they admired so much into a counterestablishment figure who always identified with the little man and whose own approach to science would surely, they believed, have led him to see what was really there. For them, and for many since, Douglass Houghton has ceased to be a person and has become instead a symbol of true, empirical science and of the accessibility of the secrets of nature to those with clear and unbiased vision. Despite its dubious factual underpinnings, that symbol has proven attractive enough to have retained its force for almost a century and a half.

Douglass Houghton was unquestionably the dominant figure in the development of the district during the critical years of the 1840s but, dying at the midpoint of that decade, he did not even suspect what the future would hold as far as the nature of the copper was concerned. He truly was, as he has been called, the Columbus of the district, and like Columbus, Houghton never understood the true nature of what it was that he had found. On the fundamental question of the value of the native copper of the Keweenaw, his views were decidedly conservative and very much in accord with accepted ways of thinking. His ties to the past were, indeed, at least as strong as was his vision of the future.

7

Charles T. Jackson and the Federal Survey

After Houghton's death the geological survey of Michigan entered a period of confusion, lost opportunity, and even failure. The final report, so long anticipated as the capstone of Houghton's efforts, was never to appear. At least some of the problems encountered, however, were already in the making well before his death. The financial straits of the survey were continuous, but Houghton's personal horizons had also been broadening and further political office, perhaps the governorship, was beginning to look like a real possibility. Always one with wide-ranging interests, he was perhaps spreading himself a little thin. The final report may have come to have a lesser priority in his mind, though it was to have been the culmination of the entire survey. Although the judgment that, after his death, "not one line could be found" of the anticipated document seems severe,[1] it nevertheless is clear that, even after years of work, it was still nowhere near completion.

After Houghton's death it was soon realized that some effort should be directed toward using the great amount of data that had accumulated since the beginning of the survey in 1837. Several attempts were made to do so, but none was particularly successful. Most of the effort was devoted to trying to organize the results of the combined linear and geological surveys in the Upper Peninsula into some sort of meaningful form. William Burt and Bela Hubbard had been Houghton's primary assistants in that fieldwork. They were each asked by the administrators of Houghton's estate to assemble a report on the region in which they had been working to send to Lucius Lyon, the surveyor general. "It would not be expected," the administrators wrote, evidently trying to relieve their minds, "that the information contained in these reports would be as complete and as accurate in detail as it would have been could they have been prepared by Dr. Houghton himself."[2]

These reports subsequently appeared in two publications in 1846, first in a small booklet coauthored by Jacob Houghton, Jr., Houghton's brother, and T. W. Bristol, and shortly thereafter in an expanded form by Jacob Houghton, Jr., alone.[3] They were intended to provide the public, particularly potential investors in the copper mining efforts beginning to blossom, with accurate information on the mineral region. They were basically summaries of work already done. For interpretations, each depended largely on the earlier views of Houghton, and they also included substantial excerpts from his copper report. Hubbard, in a report of his that was included, regretted the contrast of his effort with what would have been done by the "master mind," the "genius," the "philosophic mind" who was now gone. Burt and Hubbard continued their work on the combined surveys during the 1846 season, and although these reports helped distribute more widely what was known of the mineral district, they were a far cry from anything like the final report that had originally been anticipated for the entire state.

In his annual message in January 1846, Governor Alpheus Felch bemoaned the "melancholy disposition of Providence" in the death of "the faithful and efficient officer" of the survey. He noted that because no person had been authorized to complete the final report, its very appearance was threatened. However, he also noted that the "embarrassed condition of the treasury" meant that funds could be made available only for expenditures that were "absolutely indispensable."[4] At the same time, in his report to the state legislature, topographer S. W. Higgins urged some sort of action to ensure that the survey could be brought to a successful conclusion in the light of the work that had already been done. He pointed out that all the field notes of the survey were in the hands of Bela Hubbard, many maps had been sent to the engravers, and "the whole is far advanced towards completion."[5]

A select legislative committee then took up the entire issue. The members were appreciative of the difficulties encountered, and were very much impressed with the progress made to that point, noting that the maps already engraved were perhaps the finest ever done in the country. Strongly endorsing the continuation of the combined surveys in the Upper Peninsula, they also calculated that the total cost of the survey, at a time when the state was squandering hundreds of thousands of dollars in fruitless railroad speculation, was about the cost of laying two miles of railroad track. The committee closed its report by strongly recommending that the survey be continued, with appropriate salaries for the position of state geologist and topographer. Consequently, in May the legislature adopted a detailed resolution authorizing the governor to "appoint some competent and suitable person . . . to prepare a final report upon the geology of Michigan."[6] Bela Hubbard was, without doubt, the obvious and proper choice as that person, and S. W. Higgins was likewise fully capable of completing the publication of the maps. Hubbard made a strong personal effort to get support for the project from legislators.

Nevertheless, ever penny wise and pound foolish, "in a way that did little credit to some of the wise and economical minds of our state," the legislature apparently took no action whatever.[7] Hubbard, who had been continuing the work of Houghton on the combined surveys, specifically intending to use the results to complete the final report, was greatly disappointed. Alexander Winchell, Michigan's second state geologist, wrote in 1886 that "one can not help feeling that the government of Michigan committed a crime against the people and against posterity."[8] As a result, the first geological survey of Michigan, whose final years had been virtually without any financial support, now ended in total chaos. A large collection of valuable notes, drawings, and maps was scattered, and even the possibility of any form of a final report was, at last, doomed to oblivion.

Strong support was maintained by some, however, for continuing the combined linear and geological surveys that Houghton had begun. Hubbard believed very strongly that they had already proved their worth, and he campaigned for support so that they could continue. In retrospect, it seems clear that the completion of the combined survey would have been the most efficient procedure in the long run, yielding the greatest return for the time and effort expended.

It was not to be. What was happening in the district had become part of a national revolution occurring over how the mineral lands of the United States were to be managed. Logic and fiscal efficiency may have been on the side of the combined linear and geological surveys, but the issue had now effectively passed beyond the possibility of local control. The next significant steps were to be taken not in Michigan, but in Washington.

It was clearly in the best interests of the country to develop the copper mines in an orderly fashion, but exactly how this might be accomplished had long been uncertain. That the leasing system was not working was increasingly evident to all. In the initial heat of the rush, speculation had centered on paper rather than on copper, on the trading of leases and permits. Many of these were of dubious value, and there had been little in the way of actual mining attempted. Now, with the speculative bloom fading, actual mining, with its ores, mineshafts, and surface plants, was increasingly becoming the issue of the day.

From the beginning the leasing policy of the federal government had had its opponents, and once again voices were being raised in support of the direct sale of the mineral lands. The leasing system, with its complicated procedures, its limited times of access to the land, its payment of a percentage of the metal to the government, and its great potential for corruption, was proving overly complicated and very expensive to the public treasury. The outflow of public money was beginning to cause concern because there was little prospect of any return in sight. By January 1846 over twenty thousand dollars had been expended in administering the leasing program since 1843, but not one cent had been returned to the government through the required tax on copper actually mined.[9]

In August 1845 the secretary of war had sent two agents to the copper country to evaluate the effects of prevailing governmental policies. They stayed there until the fourteenth of October, the day after Houghton's drowning, "a period quite as late as we could with safety remain," and they submitted their report the next February.[10] Their opinion was that the copper mines were "beyond all question, the richest that ever have been discovered," and that they would prove to be the source of much revenue to the government. For a variety of reasons, however, they found the current leasing policy to be "radically defective," comparing the federal government to those governments in Europe that held the land and regarded the citizens as only tenants. What prudent man, they asked, would invest the necessary capital when his possession of the land was limited to only nine years? They also concluded that the entire leasing system, which had simply been extrapolated to Lake Superior from the lead-mining districts of the Mississippi valley, was completely "without authority of law. The plan, therefore, of disposing of these lands by lease, should be abandoned." They should instead be sold outright, and "the purchaser should be left at liberty to work the mines at his own pleasure. This plan removes all the difficulties attendant upon the present system." They also reported, incidentally and contrary to much earlier speculation, that there was no lead ore whatsoever in the entire region.

The pressure for change was building. The possibility that someone might obtain a claim, valued at perhaps as much as a hundred thousand dollars, simply by asking for a permit and signing a name, was seeming more and more unfair. The government was being drawn into lawsuits caused by the conflicting claims. The entire affair was assigned to the Committee on Public Lands, which reported to the House on May 4, 1846. They found both the leasing system and the expenditure of funds to implement it to be totally without legal basis. The committee also was greatly concerned with the social implications of the reservation of mineral lands by the federal government. "Tenantry is unfavorable to freedom," they wrote. "It should be the policy of republics to multiply their freeholders, as it is the policy of monarchies to multiply tenants."[11]

The legislature of Michigan was not long in sensing the significance of what was happening in Washington. On the thirteenth of April both houses of the state legislature passed a resolution and ordered that it be communicated to Congress. It declared that all leases of the mineral lands in Michigan were "contrary to the interests and policy of this state, in contravention of the acts of Congress . . . and an unauthorized exercise of power."[12] Michigan saw an opportunity to try to seize control of its lands and benefit from their sale. A state senate committee backed up the claim with elaborate arguments based on states rights as opposed to those of the federal government. It concluded that "your committee would recommend the full and unreserved assertion, upon the statute book, of the sovereign right of Michigan to all mines of precious metals within her borders."[13] On May 6, 1846, the issuing of further leases was halted.

The outright sale of the mineral lands was clearly the procedure of choice now, for which there was considerable agitation from many sources. Citizens' groups petitioned Congress for action. Business interests were supportive. A bill to that effect was introduced in Congress. Yet, those favoring the policies of President Polk prevented action being taken. Factional politics clearly seemed to be playing a role in the delay that helped prevent the emergence of a coherent governmental policy toward the district. The industrial enterprise of the North, it seemed, could not escape "the heavy hand of southern oppression," and for a year the situation remained unsettled.[14]

This unclear governmental policy probably contributed to a more general disillusionment that was beginning to set in as the force of the copper rush was being dissipated. Speculative copper stocks declined dramatically in price and enthusiasm began to fade. Reality, in the form of the realization that copper mining was a long-term, risky, and expensive business, one in which the possibility of a quick fortune was very unlikely, was beginning to set in. Finally, on December 8, 1846, President Polk recommended the sale of the lands, but, to a great extent, the intense "boom phase" of the copper rush was over.

There remained, however, a fundamental prerequisite to any such sale. Much of the land had still not been accurately surveyed, nor had the lands likely to be mineral-bearing been clearly delineated, and a survey was an absolute necessity before such properties could be sold and the many conflicting claims resolved. Hubbard, Burt, Lyon, and all those with the interests of Michigan at heart insisted that the efforts of the state, based on the combined linear and geological surveys organized by Houghton, were adequate to the task. Further, they argued, the publication of the results of the surveys of 1845 demonstrated that much progress had already been made toward the goal.

These claims are undoubtedly accurate. Previously, the federal government had explicitly approved and paid for Houghton's work. It was designed to do exactly what was required: accurately survey the land while producing the needed information on the distribution of those areas likely to be valuable for their mineral deposits. Having finally decided to make the mineral lands available for sale, however, the powers in Washington evidently did not wish to lose what direct control they had over the future of the district. It was therefore decided to abandon the proven combined surveys that had been largely under the control and direction of the Michigan people. Instead, a separate, federal geological survey was initiated that was to be conducted by federal appointees and was to be independent of the state linear survey that was subdividing the lands.

On March 1, 1847, Congress passed *An act to establish a new land district, and to provide for the sale of mineral lands in the State of Michigan.* The provisions of the act called for the secretary of the treasury to cause a geological examination and survey of the "Lake Superior land district" to be made, before the lands were offered for sale, with the rights of leaseholders of record being protected to allow efforts in progress to continue.[15] Shortly thereafter, Treasury Secretary Robert J. Walker, in compliance with the act, made his appointment to fill the position of United States Geologist, the head of the federal survey. His choice, which was roundly condemned in Michigan and which was to bring the resulting survey to the very brink of disaster, was a physician, chemist, and mineralogist from Boston named Charles T. Jackson.

Charles T. Jackson

Charles Thomas Jackson (1805–80), like Houghton, was a physician who came to play a significant role in American geology in the first half of the nineteenth century (fig. 17).[16] He came from an old New England family, was a brother-in-law of Ralph Waldo Emerson, and graduated from Harvard medical school in 1829, winning the Boylston prize for his dissertation. In the same year he went to Europe, studied medicine and geology at various French institutions, and traveled extensively, making many observations of the geology of the continent. He also established friendships with prominent European medical and scientific figures.

His publications reveal a great variety of scientific interests. Early in his life he had cultivated a strong geological and mineralogical leaning, and even before completing his medical degree he traveled to Nova Scotia with his friend Francis Alger to study its geology. The traprock with some minor amounts of native copper that they encountered there resembled in many ways the geology he was later to find in the Keweenaw Peninsula. Their observations in Nova Scotia formed the basis of Jackson's first published geological work which gave him,

Fig. 17. Charles T. Jackson, U. S. federal geologist in the Lake Superior region from 1847 to 1849. From Merrill (1924).

early in his career, something of a reputation as a mineralogist. Also strongly interested in chemistry, he gradually abandoned the medical profession and devoted increasingly more of his efforts to his scientific pursuits.

Jackson played an active role in the state geological survey movement, and served successively as the first state geologist of Maine, Rhode Island, and New Hampshire between the years 1836 and 1844. In contrast to Houghton, he produced his final reports promptly, and perhaps because of his chemical background he leaned strongly toward mineralogical considerations. In his surveys he was always looking for the practical and economic application of what he saw, an ideal characteristic from the perspective of the state legislatures that employed him.

It has been pointed out that both Schoolcraft and Houghton professed scientific philosophies that emphasized direct observation and facts, and minimized, even demeaned, theories and speculation (although, in Houghton's case, profession and practice differed considerably). In that philosophy they were one with Jackson, and perhaps one with the professed general scientific spirit of the age. The proper scientist, said Jackson, "prefers to allow the facts to remain unexplained, rather than bedim his vision by the colored glasses of theory, and be thus led to false conclusions. I value more discrepant facts than a general theory—holding that hasty generalizations have done much to obscure the vision of the scientific world."[17] And, most succinctly, "I only hold the pen; Nature dictates the facts."[18]

There was, however, more to Charles T. Jackson than this. "He had a well-trained, well-stocked, well-disciplined brain, but there was a sick spot in it." During his long life he managed to become one of the more controversial figures in nineteenth century American science, and in American life generally. He is most often remembered today for his involvement in several frequently bitter priority conflicts that left their marks on the scientific and social scenes of his day. A consensus emerged that did not enhance Jackson's personal or scientific reputations, and some of the judgments returned against him were, and are, very severe. These have made evaluating Jackson's true scientific accomplishments problematic, even today.[19]

Because he was directly involved in one of these controversies when he was appointed to the federal survey, and because he also subsequently made a definite but little-noted priority claim for himself as an outgrowth of that survey work, a brief examination of Jackson's personality and these important episodes in his life is appropriate here. The conflicts all followed the same predictable pattern: a discovery would be made and announced by someone, Jackson would subsequently declare that true credit for the discovery belonged to him, and a controversy would ensue. At least four such incidents in which he was involved have received attention in the past: the invention of guncotton, the digestive action of the stomach, the invention of the telegraph, and the discovery of the anesthetic effect of ether.

The Priority Conflicts

In the first mentioned case, in 1846 Jackson claimed to be the discoverer of guncotton, a powerful new explosive, after an earlier announcement to that effect by C. F. Schönbein.[20] The second involved the experiments done in the 1820s and 1830s by William Beaumont, a physician stationed at Fort

Mackinac where a young Frenchman received a gunshot wound in the stomach that did not heal completely. Beaumont then conducted a long series of pioneering studies on digestion through the opening that remained. In 1834 Beaumont was in Boston where he gave Jackson a vial of gastric juice to study. When Beaumont, along with his subject, was then transferred to a western post, Jackson immediately tried to forestall the transfer and carried a petition to the secretary of war, arguing that the presence of Beaumont was necessary in the Boston area. This would permit "an eminent chemist of Boston" (himself) to take advantage of "a rare and fortunate opportunity of demonstrating important principles in physiology" because the opportunity, "if neglected, will be lost to our country forever."[21] Beaumont, however, had already published many of his observations in 1822 and 1833, which had become widely known and which have earned him a permanent place in the development of the understanding of the nature of digestion. The attempted delay did not occur, but Jackson's self-serving actions were a foreshadowing of what was to come in other areas.

The third case concerned the invention of the telegraph. On the ship on which Jackson returned from Europe in 1832 a group of passengers, which included Jackson and Samuel F. B. Morse, had some conversations about electricity and the possibility of communicating by electric signals. Five years later Morse contacted Jackson to tell him of the success of his telegraph. Morse was surprised and disturbed at the reply, in which Jackson referred to "our" instrument. This initiated the beginning of a campaign that Jackson waged trying to win for himself recognition for inventing the telegraph. He was later to claim that on the ship Morse had no concept of what electromagnetism was until Jackson elucidated the subject for him, including the principles of the telegraph itself. The affair degenerated into claim and counterclaim, with the case finally being clearly decided for Morse.

The author of a contemporary analysis of the episode wrote of Jackson that he, "according to his own story, was first a *mutual* inventor, then a *principal* inventor, and finally the *exclusive* inventor!" He concluded that Jackson was a "scientific monomaniac. . . . Giving due weight to the facts herein established, who can rely on Dr. Jackson for the truth in any matter of science or exploration? His vanity amounts to insanity."[22] And the next episode was even more controversial, being, according to the author of a recent major study of American science, "a classic instance of the lust for scientific priority run mad."[23]

In October 1846, William Morton successfully administered ether as an anaesthetic to a patient who then underwent a surgical operation without pain. Morton, a dentist, thereby initiated both a revolution in man's age-long battle against pain and a priority dispute that was to rage for many years, involving himself, Jackson, and Horace Wells, a former partner of Morton. Jackson apparently proposed the use of ether to Morton, and controversy then flared over the significance of his role. The entire affair was a sordid one. Jackson characteristically insisted that the whole concept was his, and that Morton had simply been his agent, acting under his specific directions, in the successful demonstration. The controversy raged, congressional hearings were held, and as it became clear that the judgment of the time was going to fall to Morton, Jackson once again saw his attempt to establish himself as a benefactor of mankind slipping away.

Jackson's subsequent conduct has brought on him judgments that are sometimes very harsh. According to one analyst, Jackson waged against Morton "one of the most horrifying vendettas ever recorded among civilized men." Jackson was a "psychological assassin, a sadist and a paranoid psychopath who found it congenitally impossible to give credit where credit was due in any sphere of activity. . . . Jackson was always thrusting himself forward as the originator of other men's inventions or discoveries. It was probably a deep-down consciousness of inadequacy which devoured him and compelled him to claim other people's accomplishments as his own."[24] Another author sadly concluded, "What had once been a small sick spot in Charles T. Jackson's otherwise vigorous and healthy mind finally claimed the whole of it." The last seven years of his life, his mind gone, were spent in an institution for the insane.

This, then, was the man chosen to head the new geological survey of the Lake Superior land district. For many in Michigan the survey was now thrice-cursed. They had not wanted it, preferring instead the combined surveys of Houghton; it was now to be headed by an outsider rather than someone from the state; and finally, that someone was Charles T. Jackson. The Michigan delegation in Congress, perhaps with the prodding of Hubbard, fought valiantly. They tried to change the appropriation bill to require the head of the survey to be a resident of Michigan, to have all of the chemical work on the minerals done at Detroit, and to have all assistants be citizens of Michigan, but it was to no avail.[25] Politics clearly played a role in what was happening, and Jackson had friends of high standing in Boston who used their influence to have him chosen. "The appointment of Dr. Jackson to the post of U. S. Geologist over that region," Hubbard later lamented, "unexpectedly and secretly made, at once put an end to the prospect. But for the change in administration, this state of things would have been amended."[26] For better or for worse, it was to be an independent geological survey, and it was to be headed by Jackson.

The Federal Survey Fiasco

It was for worse. The plan had been to make an accurate mineralogical examination of the lands to evaluate their potential for mining and to analyze those mining efforts already in progress. As it turned out, the survey under Jackson was almost a complete fiasco, saved only because his two principal assistants, John W. Foster and Josiah D. Whitney, were genuinely capable men, and several other assistants had served with Houghton and were familiar with the region.[27] Whitney, later to become a prominent geologist in his own right, had been something of a student of Jackson, who guided him into his career and who had employed him as an assistant in the New Hampshire survey. Jackson himself was no stranger to the Keweenaw, having previously been to the district several times, once with Whitney as an assistant. He had been a consultant to several early mining companies at the height of the rush, and he thereby played a role in generating the earlier copper fever in the East. His leadership of the survey, however, was to prove to be a disaster.

The survey took to the field in the spring of 1847.[28] They reached Copper Harbor on the twenty-sixth of June, and much of the time the party was divided into three groups, one under Whitney, a second under John Locke, and the third under Jackson. Serious mining efforts were beginning all up and down the district, and time was spent examining in considerable detail the mining properties being opened with varying degrees of success, including the fabulous Cliff location. Jackson's strong mineralogical interests were clearly evident in his descriptions of the mines. After visiting Isle Royale, Jackson returned to Boston on the third of October and spent the winter, so he later insisted, in "analytical and metallurgical researches, applicable to the working of the metals and ores of Lake Superior."[29]

For the 1848 season, John W. Foster replaced John Locke, and this made Foster and Whitney Jackson's primary assistants. Once more Jackson assigned each of them to head a group working on an area. Their individual reports, and those of Jackson, while they do exhibit some evidence of personality clashes and disagreements about the conduct of the survey, in themselves provide little evidence for what was apparently happening. However, it was becoming increasingly evident that Jackson's leadership role as chief geologist of the survey was disintegrating and that a crisis was looming.

Even at the beginning of the field season of 1848 the problems were becoming more obvious. Appropriations were held up by the continuing attempts of the Michigan delegation in Congress to modify the conditions of the survey, or to again connect the geological survey with the linear survey that was continuing on the old pattern.[30] Lyon, Hubbard, Burt, and Higgins all sent letters to Washington dated between April 24 and 28, 1848, which were therefore clearly part of a concerted effort. All of them complained about Jackson's work.

Lyon denounced Jackson's efforts as "uselessly extravagant," and charged him with appropriating the work of Higgins, Hubbard, and Houghton without acknowledgment (a charge that still has some currency). Burt declared that the federal survey was of no value. Hubbard reinforced the plagiarism accusation: "But this willful ignorance and self-conceit might be excused, did not Dr. Jackson show also a determination to make use of the labors of others, while he appropriated the credit of them for himself." Higgins likewise wrote that he "cannot conceive of a more ridiculous attempt of any man thus to depreciate and afterwards to appropriate to himself the labors of others. Instead of being greeted as was Houghton, disgust and dislike met him everywhere." Higgins further listed the names of thirteen influential citizens of Michigan from whom similar letters might soon be expected. The pressure was building.

Robert Walker, the secretary of the treasury who had originally appointed Jackson, refused to heed the complaints.[31] Jackson himself journeyed to Washington and managed to get his appropriated funds, but the party was delayed in getting to Copper Harbor until the eighth of July, well into the working season. On the eighteenth he wrote a long letter from Fort Wilkins to Walker, which dealt mostly with the opposition that the federal survey was receiving.[32] Jackson then passed to the issue that was to reappear in the future. There was the distinct feeling among many in Michigan that Jackson was claiming for himself discoveries that Houghton had made. But, he protested, "it has always struck me as a very improper procedure on the part of Mr. Hubbard and his friends to *attempt* to make an unpleasant issue between the labors of my late friend Dr. Houghton and myself, and that they should do this for selfish ends shocks every one who knows the facts." The issue would not disappear, and it emerged again as part of the events that followed Jackson's fall from grace as the survey degenerated.

Foster and Whitney became increasingly disenchanted as the season progressed, and they began to entertain serious doubts about their ability to continue as part of the survey the next year. They learned that the Michigan people were determined to have Jackson removed, and they feared that they would be implicated as "aiders and abettors of his misconduct."[33] Their position was difficult. They had no desire to make a public affair of their concerns. On the other hand, they also knew that they, as subordinates, were required by law to report any delinquencies by a superior or else possibly be considered accessories to them. They compiled a long list of their complaints, and even a partial recitation would seem to be a caricature of reality, were it not fully backed by the direct testimony of reasonable men.[34]

They charged: that Jackson spent less than four months each season in the field; that in the fifteen months in Boston, his lab work did not total fifteen days; that during the field seasons he took money on the side for private consulting work on the mines and used government funds to pay for the private analyses; that he filled subordinate positions with his unqualified pupils from the East; that he did not complete a single assay for the government; that he made no effort to complete the required survey report; that he did not travel over thirty miles in the mineral region, and then only "over blazed and beaten trails"; that as a result of "his aversion to labor and his disposition to loiter about the mines" he did not make "a single observation, in the execution of the work; and yet, when completed, he has been unwilling that they should have any share of the credit"; that he "appropriated the public funds for his own use"; and so on.³⁵

Supporting evidence from others also abounded. Samuel W. Hill, one of the surveyors and a veteran of the copper district (and, apparently, of "What the Sam Hill?" fame), claimed that most of Jackson's time "was spent in reading works of fiction and talking about the Ether Controversy." One could not, he opined, go on listening to him for weeks "without believing him positively insane." Charles Joy, who worked in Jackson's laboratory, said that all discussion centered on the ether controversy, and that everyone became thoroughly disgusted with it. Jackson was also evidently having something of a drinking problem. "Is he not," Joy asked, "partially deranged?" Articles were even appearing in newspapers raising questions about the quality of the work Jackson was doing.³⁶

Clearly, something had to be done. Foster and Whitney decided that their only choice was to submit their resignations. This required them to state openly their reasons for doing so, to avoid the countercharge that they had abandoned the survey in midstream. Their intentions, and their list of charges, were submitted to Jackson by a friend in April 1849. Foster then visited Jackson and advised him to resign, for "all the assistants would be summoned to Washington to swear to various particulars, which they were prepared to do."³⁷

Jackson, finding himself cornered, blustered and denied everything. A group of four of his friends, including Dr. Augustus. A. Gould, a distinguished physician and naturalist well known for his work in conchology, and George L. Ward, a prominent Bostonian instrumental in Jackson's original appointment to the survey, reluctantly agreed to be an arbitrating committee. Upon review of the evidence, they were "unanimously of the opinion, that Dr. Jackson should unconditionally resign," and that the survey should be turned over to Foster and Whitney for completion.

A document was then drafted that required the three to say nothing whatever that might be detrimental to the public reputation of the others but to "sustain and uphold them in all honorable ways." It was signed by the three principals and the four members of the committee, and on the twelfth of April Jackson submitted his resignation in writing to Thomas Ewing, the secretary of the interior, for reasons both "private and professional. I take the liberty to recommend that the completion of the survey be left to my assistants." The Jackson episode, at last, was over. Or so it seemed.

Within a few days, Jackson's fire returned. Friends who had not heard the details of the charges against him urged him to reconsider. Claiming that his resignation had been extracted from him under duress, he dashed off a quick note to Secretary Ewing, asking him to "please not accept my resignation until you have an explanation from me." To Foster and Whitney, he wrote that "I would inform you that I hereby rescind all agreements between you and myself relating to my resignation of my office of U. S. Geologist, and I hereby notify you that I accept your resignations of the appointments that I had conferred upon you as my assistant geologists."[38] He then demanded a full examination of the charges against him. Augustus Gould, his friend who had worked so hard to get the original agreement, was dismayed; "he will come out of this little less than ruined. His manner of conducting his affairs is such that he would spoil even a good cause."[39] The case proceeded to a full-scale investigation in Washington.

Trying to clarify the situation, the clerk of surveys wrote a summary of the case to that point. Its opening words exhibited, perhaps unintentionally, the farce into which the whole affair was degenerating:

> This case stands thus: Dr. Jackson resigned as principal geologist, but has since asked permission to withdraw his resignation. Messrs. Foster and Whitney, principal assistants, have also resigned, to take effect on the 30th May next; but in the meantime, they have been discharged by Dr. Jackson, when, however, his own resignation was in abeyance, he had not the right to act in the matter. The controversy appears to be such, that the principal or assistant geologists will have to quit the service, as it is not perceived how they can harmoniously co-operate, after the circumstances which have transpired, and the allegations which have been made.[40]

Jackson produced a letter containing "one hundred refuting charges," in which he defended himself and argued that he should be allowed to finish the survey. Foster and Whitney countered with a lengthy reply. "We strenuously maintain that we have not sought to become the public accusers of Dr. Jackson," they wrote, "and that we have taken no steps in this matter but such as we were compelled to, by a just regard for our own reputations and the public interest."

Their reply was an excruciatingly detailed document specifically directed, point by point, against Jackson's attempted defense.[41] Their charge, that except for his visits to the mines and boat trips Jackson had not traveled over thirty miles in the district, was supported with a map on which "the yellow lines represent the points he visited the first year. The blue lines, those visited by him during the second. . . . He did not determine the geology of a single township, or a single quarter section." He "remained in camp and amused himself in fishing." He "spent his time in a bar-room, at the mouth of the river, haranguing miners, discussing and administering ether for the amusement of the crowd." He "was deeply absorbed in reading *Ten Thousand a Year*. He very soon discovered a marked resemblance between Tittlebat Titmouse, the hero of that work, and Dr. W. T. G. Morton, the rival claimant for the discovery of etherization." His enthusiasm at identifying characters in the fiction he read "became so annoying, so interfered with the progress of the work, that they finally had to request him, in decided but respectful terms, not to intrude on them in business hours." Of the many minerals obtained, "almost every really valuable specimen collected in the progress of the survey, is to be found in the private cabinet of Dr. Jackson, while the Government possesses a mere accumulation of rubbish." He repeatedly appropriated the work of others for himself, and presented their results as his own discoveries. And so on, and on, and on.

The entire matter was examined by the secretary of the interior and the attorney general. Time was wasting, as the intent had been to offer the mineral lands for sale the coming June. Further delay could not be afforded and the result was inevitable. On May 16, 1849, Foster and Whitney were informed that "the labors in the field of Dr. Jackson, U. S. Geologist for the Lake Superior District in Michigan, shall be considered as closed." Jackson, as part of his termination, was ordered to produce a report before the first of the next November summarizing the survey to that point, and Foster and Whitney were instructed to "go on and close up all further operations in the field, that may be necessary for a full, practical, and scientific report . . . necessary to a proper execution of the act of 1st March, 1847."

The affair was not over for Foster and Whitney, however. Jackson bitterly attacked them both, and made a vicious and slanderous attempt to exclude Whitney from membership in the American Academy of Arts and Sciences. The academy then held a hearing on the unfortunate affair before its fellows, and the results were "fatal to Dr. Jackson's reputation, not only as a man of science, but as a gentleman." A resolution was proposed to expel Jackson, but out of compassion it was never publicly presented for consideration.[42]

As far as the copper district was concerned, the Jackson phase of the federal survey was finally concluded. Foster and Whitney, now free to do what they had always known should be done, moved quickly to complete the survey and to produce the report that, after it was published in 1850, was recognized for the next thirty years as *the* authoritative treatment of the geology of the copper district.[43]

And yet, Jackson had one last card to play. On November 10, 1849, only nine days past the deadline he had been given, he officially ended his role in the survey by delivering his report to the secretary of the interior. It was a heterogeneous, lengthy, confusing document, composed largely of fragments written by his assistants that he had thrown together and for which he had written a disorganized introduction. It was superseded almost immediately by the Foster and Whitney report of the next year. However, what may be the most interesting item in the report came after his historical review of the Keweenaw district and of those who had contributed to an understanding of it. It was a specific discovery for which he claimed priority, and the claim bore directly on the mineralogy of the district and the question of the true nature of its copper.

8

JACKSON AND THE EARLY MINING EFFORTS

To conclude his association with the survey and to prepare the closing report that was required, Jackson asked all his assistants to submit their findings to him. Even Foster and Whitney were directed to provide him with all the results of their investigations to that point. In the notes that Whitney submitted he could not resist taking the opportunity to snipe at Jackson, who would then respond in kind in brief editorial footnotes. Whitney: "Dr. Jackson, who is making a picturesque tour round the lake on board the *Julia Palmer*." Jackson: "This is a gross misrepresentation." And so on. Jackson nevertheless did little editing, as he must have realized that his report would soon be eclipsed by the final work of his former assistants. As a result, Jackson's report was in large degree simply "one thing after another." It exhibited little coherence, and even included the old combined survey reports of Burt and Hubbard that had been completed in 1845–46. It was published in December 1849 as Senate Executive Document Number One and as House Executive Document Number Five. By the time all the subreports of the assistants were added, it ran to 564 pages.[1]

The Jackson Claim

As had become virtually standard practice in such reports, Jackson began by providing in a dozen or so pages a fairly comprehensive historical overview of the copper district. "Justice to our predecessors," he stated, required that their accomplishments be fairly noted. First describing the early efforts of the Indians to use the copper, he then proceeded to the French, beginning with an account published in Paris in 1636 by, as he had it, one "Lagarde," by which he undoubtedly meant Gabriel Sagard, discussed in chapter one.[2] He briefly noted the individuals who had called attention to the copper down through the years, culminating with the work of Houghton. Passing to the events that led to his own arrival on the scene, he explicitly claimed priority for discovering the uniqueness of the native copper.

It has long been known that native copper is occasionally found in trap rocks, masses many pounds in weight having been discovered in those rocks, in Connecticut, New Jersey, and Nova Scotia; but no one then supposed that the occurrence of such masses of this metal was of any practical interest, or that it was even probable that any locality would be discovered where it could be economically mined. It was generally believed, by geologists, mineralogists, and miners, that the few isolated lumps of native copper which were occasionally found . . . were accidental, and were derived from the fusion [melting] and reduction of pieces of copper ores which were originally included in the partially altered or fused rocks.

In this state of scientific opinion it might have been hazardous to the reputation of any geologist, who was not prepared to demonstrate the fact, to declare his belief in the practicability of mining with advantage for *native copper*. This I was fully aware of.

It is well known that the results of my examination of the mineral lands on Keweenaw point were the establishment of the fact that *native copper and native silver* existed there *in regular veins, which could be advantageously wrought by mining operations*.

[I] then explored other veins, and pointed out the superiority of the *metallic lodes* over that of the *ores of copper* —a result quite contrary to the general opinion of miners and geologists, but which has been most fully sustained by subsequent experience. Experience has since proved that the views I then promulgated were correct.

It is due to the country and to myself that this exposition of the facts should be published . . . trusting thus to place them beyond the reach of any subsequent explorer who might set up claims to them [all Jackson's emphasis].[3]

What is one to make of this remarkable claim? Surely it bears an amazing resemblance to the earlier claims that he made for himself, none of which has been acknowledged in his favor and all of which have seriously compromised his scientific reputation. The concern for his priority, the pathetic need to be recognized, the fear that what is due him may be usurped by another, all this is vintage Jackson; all this has been seen before. He made a similar effort to establish his priority at the 1849 meeting of the American Association for the Advancement of Science. In a presentation to that group that was published the next year in the *American Journal of Science*, he insisted that he deserved the credit for predicting, contrary to all geological opinion of the time, that native copper would be profitably mined, and "the predictions I then made are now fully verified."[4]

And yet again, trying to refute the charges of Foster and Whitney in the messy affair that led to his dismissal from the survey, he echoed the theme:

> Having, on my own responsibility, and against the opinion of scientific men in this country, and practical miners, FIRST pronounced the working of veins of metallic copper to be feasible and profitable, and having FIRST *demonstrated the fact* now realized by practical operations on Lake Superior, that those mines would prove richer as they descended into the rock, I naturally felt a peculiar pleasure in presenting to this country and to Europe the facts relating to a district which presents an entirely new chapter in the history of metallurgy [his emphasis].[5]

As he had so many other times, he was once again in full pursuit of credit for a discovery, trying to stake out a place in the history of the district. He insisted on pressing this claim before any and all who would listen, and undoubtedly some who wouldn't. Others had heard this kind of refrain from him too many times before, however, and were not about to concede him anything.

There was at this time virtually unanimous agreement that the honor of such discoveries belonged to Houghton and not to Jackson. Houghton had described the copper-bearing veins much earlier, and Foster and Whitney did not miss the opportunity to express their displeasure at this infringement by Jackson on what they believed was a result of the work of Houghton and other earlier investigators. They said that they were at a loss to know why Jackson would claim this as his discovery; years before Houghton had gone over the same country in great detail. They further insisted that what he was claiming as discoveries were the general opinions of most people searching the region, and that Jackson had been "simply borne along with the mass." They concluded their dismissal of his case with a statement of their philosophy of geology, which strikes a curious note given the nature of the debate: "Geology is eminently a science of observation," they wrote. "Mere theorizing, or an ingenious adaptation of particular theories to particular phenomena, avail not."[6] Clearly, then, at least as far as Foster and Whitney were concerned, this claim merited no more consideration than earlier ones that had already earned for Jackson such severe denunciation.

In fact, down through the years Jackson has been conceded at least some role in the opening of the region to mining. It is now generally agreed that his reports, especially those to the trustees of the Lake Superior Company, helped raise the copper fever in the East to a white heat and to promote the wild speculation in the Michigan mines that followed. He may well have been, contrary to the implication of Foster and Whitney, a major influence in directing eastern capital toward Michigan. This much is now generally acknowledged.

At the same time, however, his work is almost always interpreted today as having been supportive of, and built upon, the earlier and more original work of Houghton. In one analysis of the copper rush period, for example, the author states that Jackson's mining report of 1844 "received wide publicity; it was the confirmation of Houghton's 'backwoods geology' for which eastern businessmen had been waiting."[7] Is there any reason, then, to believe that Jackson's claim has any more merit than those of the earlier controversies in which he was involved? Was he trying to steal the credit that belonged to Houghton, as many declared he was?

If Jackson's claims are examined closely, it is evident that in none of them does he consider Houghton a possible contender or rival. Houghton, it appears, may have recommended Jackson to Henshaw, which led to Jackson's further involvement with the district.[8] In his *American Journal of Science* article of 1845, written when Houghton was still alive, Jackson freely acknowledged the importance of Houghton's "interesting annual reports," and he anticipated Houghton's final report as a "full and detailed description of the mines and minerals" based on his "extensive and arduous surveys."[9] Jackson wrote Houghton's obituary and eulogy in the *American*

Journal of Science. In later writings he repeatedly referred to him as "my late friend." In the 1849 report, immediately before making the first claim for himself cited above, he described Houghton's work: "The first proper scientific explorations in the mineral lands of Lake Superior were made by my late friend Dr. Douglas[s] Houghton. . . . There cannot be a doubt that he was aware of the localities of most of the exposed copper veins, for he had traversed the region where they occur, and was a good observer."[10]

It seems clear, then, that Jackson was thinking of *future* possible claimants, not past, and he did not see Houghton as a rival because he understood, contrary to what the emerging Houghton tradition was maintaining, that Houghton never did advocate the value of native copper for mining. The generally accepted view today is that Jackson's opinions on the native copper were simply the "confirmation" of the earlier views of Houghton and not original in their own right but, as we have seen, this conclusion is based on a misinterpretation of what Houghton's actual views were.

But what, if anything, does Jackson deserve credit for? Or perhaps we should ask an even broader question. Was recognizing the importance of the native copper of the Keweenaw really so difficult that it should be regarded as a discovery at all? Wasn't that simple fact obvious to anyone willing to look and see what was there?

Perhaps frustratingly, the answer to this last question is both yes and no. Yes, it was obvious to some, particularly to those who represented the amateur approach to nature, but that was because they didn't know any better! On the other hand, to the more professional geologists and skilled miners of that day it was *not* obvious, because they "knew" that native copper was never found in amounts great enough to mine. They also knew that in other districts native copper, by itself, could even be an unfavorable sign for mining, because it meant that the copper-bearing minerals from which the native copper had formed had been weathered away and depleted of the metal.

Therefore, and this is very ironic, when native copper was found by those who already "understood" what copper mining was all about, what they *hoped* would happen is that it would disappear as digging went deeper, and that greater quantities of other copper-bearing minerals would be found at depth. Only in this light can we understand comments such as those made in 1845 by Andrew Gray, the assistant superintendent of the mineral district. A report of the observations he had made in the Keweenaw that summer was submitted to the House of Representatives. In it he expressed the nagging and persistent concern that many had about the great amount of native copper that was being found in the Keweenaw, and he expressed confidence that the Lake Superior district would soon turn out to be just like the other mining regions of the world, where other copper-bearing minerals are found in abundance:

> It is supposed by many that so much native copper, and in such large masses as are here found, present bad indications; and fear on the part of some has been expressed, that capital invested is likely to be lost.
>
> Large pieces of native metal have been found in all the principal metalliferous regions of the world; and upon comparison with the rich mines of the Cordilleras of South America, the mines of Great Britain, and others, the masses of native metal there found bear nearly a similar proportion to the richness and value of the ores along with which they were associated as do the masses of Lake Superior to the ores with which they have been assimilated.[11]

Anyone who might doubt that presuppositions can literally control what is observed in nature would do well to consider the second paragraph of this quotation. Despite the appearance of being nothing more than a simple statement of what had been observed on Lake Superior, it is in fact totally false: there is no similarity whatsoever in the proportion of the native copper to the other ores, for those other ores are almost completely absent. Such comments forcefully demonstrate the lengths to which persons would go in trying to convince themselves that the Keweenaw was just like all the other districts of the world, and that the native copper could be ignored in searching for other important copper-bearing minerals.

It is against this background, then, that Jackson's insistence that he had made a significant discovery must be viewed. To support that claim, he referred not to work he had done on the federal survey, but to reports he had published some years earlier when he first began to give serious attention to the district. In the summer of 1844, the year before Houghton's death, Jackson had made a visit to the copper country while employed by David Henshaw. A former secretary of the navy,

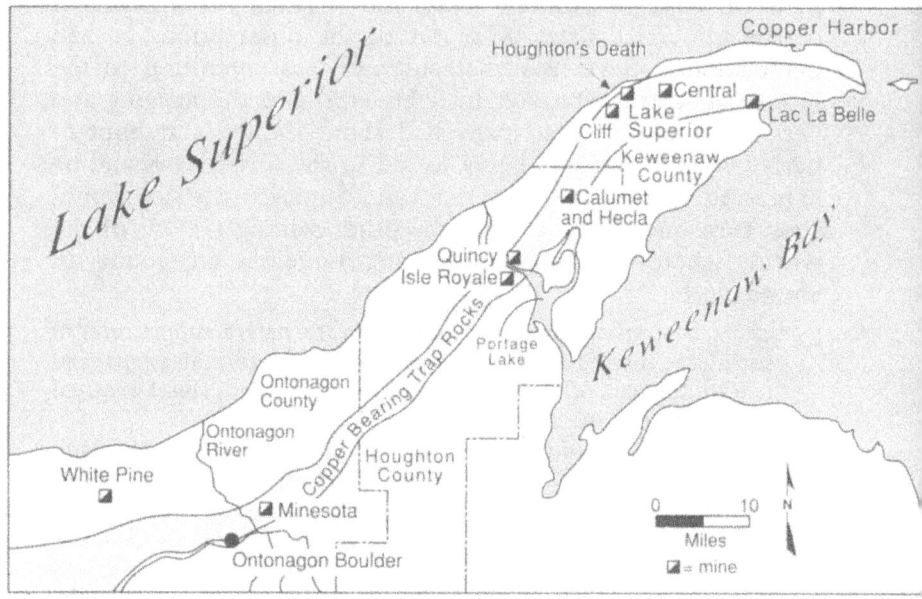

Fig. 18. Early mining locations on the Keweenaw peninsula.

Henshaw found that his governmental connections were helpful when he decided to speculate in copper leases. He was also one of the trustees of the Lake Superior Copper Company organized early in that year, which became the first to engage in systematic mining on Lake Superior. The company had acquired several mining locations on the Keweenaw (fig. 18), and in reaching them their party spent twenty-four days traveling from Boston.[12] The company's miners, who under Col. Charles H. Gratiot had already done some preliminary exploring, had built several log cabins near one of their locations on Eagle River. There, accompanied by Houghton's cousin, C. C. Douglass, Jackson engaged in a detailed examination of the veins.

He did not hesitate to declare several of them of no value, but he concluded that one of them was of great potential, "the most important and valuable of all the locations belonging to the Company yet discovered." In places the vein lay in the bed of the river, and in tracing it he found some of the rock to be an amygdaloid in which the cavities were filled with native copper and silver, along with irregular lumps and sheets of copper that created a "rich aggregate of the metals and rock."[13] He noted that where the copper appeared at the surface and had been acted upon by surficial weathering effects, a crumbling rock containing carbonate of copper had formed from the weathering of the native metal. He thereby recognized a process that was just the reverse of the traditional view: the small amounts of copper compounds at the surface had formed from what had been native copper. He was enthusiastic about the prospects, and believed that the silver found with the native copper would contribute greatly to the value of the mine.

The next year he published a lengthy article in the *American Journal of Science* titled "On the Copper and Silver of Kewenaw [sic] Point," to provide information for "the scientific community, as well as to those interested in mining."[14] After first giving an overview of the geology of the entire Lake Superior region he turned to the copper itself. He acknowledged the view that native copper is usually regarded as an unfavorable omen, and that enormous masses like the Ontonagon Boulder were "great rarities, if indeed there is another like it in the whole country."[15] Nevertheless, he believed that native copper in small masses or particles, if present in abundant amounts, might well be of value. "There are a few such localities on Kewenaw Point, and those I have examined with great attention. . . . Only two or three have been selected, by my advice, as undoubtedly valuable, and of sufficient richness for profitable mining."[16] He made a particularly astute comment about the famed "green rock," the vein of copper compounds that had long intrigued visitors to the region and on which Houghton had placed so much emphasis. He called it an "accidental ingredient" of the conglomerate rock in which it was found, and said that it was not likely to be of value for mining.

In 1845 he again was on Lake Superior, reexamining the veins of the Lake Superior Company and overseeing their mining operations. A stamping mill had been constructed at Jackson's direction by C. C. Douglass, using plans chosen from French works on mining (figs. 19 and 20). As he examined the vein further he once again made some very perceptive observations. Earlier, he had considered the possibility that the native copper had formed by the reduction effects of heat acting on preexisting ores of copper compounds present in the sandstone next to the traprock, an origin that would have been generally in accord with traditional interpretations.[17] However, the fact that the copper was found all along the vein, which ran at right angles to the contact between the sandstone and the trap, led him to abandon this theory. He reported his conclusions to the trustees of the company in November 1845 and based his comments on his work of that summer, at which time Houghton was still working in the district. Although he was, without doubt, more optimistic about the prospects of the mine than he should have been, he included some comments on the nature of the copper that were very insightful:

> Native copper, I know, is not usually found in great quantities. In Europe and in Cuba the ore is apt to change into sulphuret of copper and iron after attaining a considerable depth—but in those cases there are sufficient proofs of the secondary origin of the [native] copper from decomposed copper pyrites, while the absence of sulphurets [sulfides] and the want of iron ochre [iron oxide] both indicate the permanency of the metallic copper in your mines. I cannot see any reason to believe that there will be any change in the ores so long as there is no change in the character of the rocks in which they occur.[18]

This is a remarkable analysis, considering the time and the circumstances of its writing. Here, in brief, Jackson has set out the entire problem and its solution in a historical perspective. First, he gives a succinct statement of the traditional view based on other mining regions, with its emphasis on small quantities of native copper that originally formed as a secondary mineral from preexisting sulfide ores. Second, he notes the lack of such ores in the Keweenaw, indicating that the native copper is therefore not a secondary, relatively minor component, but a fundamentally different, primary ore mineral in its own right. Finally, he infers that the native copper could therefore continue in abundance as long as the characteristics of the surrounding rock also persist, as opposed to the traditional view that any native copper would die away at depth and give way to other, more complex ore minerals. This is undoubtedly the most astute and discerning brief statement of the mineralogy of the copper district made by any geologist of the entire copper rush period. It seems clear, then, that Jackson's view of the nature of the native copper deposits, as exhibited by his writings from 1844 and 1845, is accurate and perceptive, and that his understanding sets him distinctly apart from virtually all other geologists of that time.

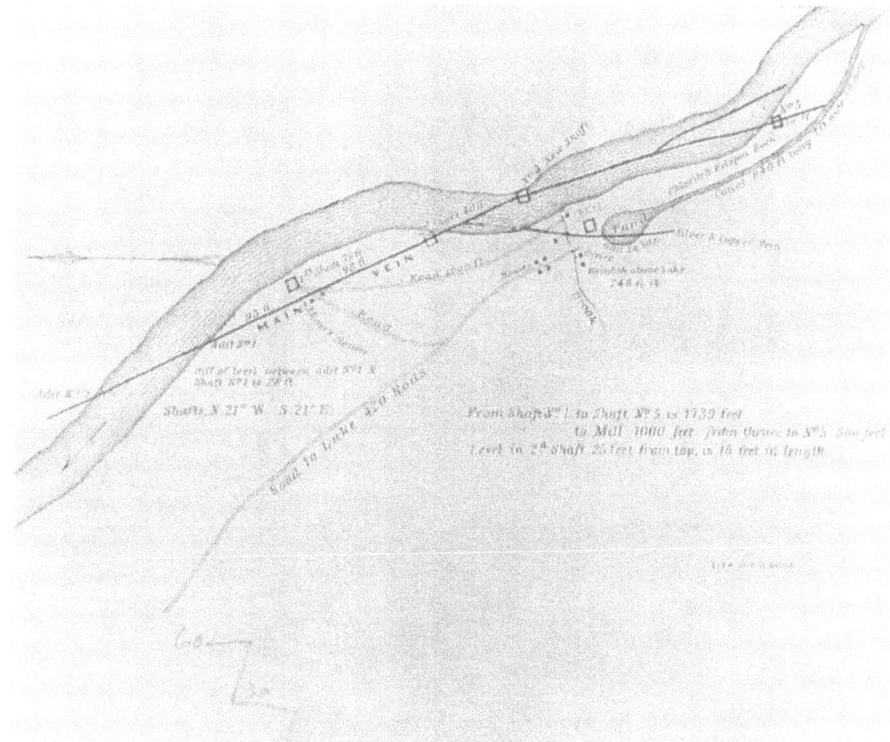

Fig. 19. *The Lake Superior Company copper vein on Eagle River, as shown in Jackson's report to the company of 1845. North is to the left. Note that in places the vein runs in the bed of the river, and that shaft number 4 is in the river itself. The square just to the north (left) of the small pond is the mill shown in figure 20. From Jackson (1845b, plate 1).*

Putting all of this together indicates that the question of the nature of the copper was in a state of considerable flux between the years 1844 and 1850. Perhaps some order can be brought to the components of the issue by separating and summarizing them as follows.

First, as early as the time of Houghton's death in 1845, certain persons who were not geologists were stating that the native copper of the Keweenaw was present in abundant amounts that might well repay mining efforts, and that this set the district apart from all other mining regions of the world. This view was the continuation of a very long amateur tradition, and it was being openly disseminated in a variety of ways.

Second, at least one of those persons, John St. John, stated explicitly that special credit for the clear recognition of this uniqueness belonged to Douglass Houghton, who, according

Fig. 20. The Lake Superior Company stamp mill, as shown in Jackson's report to the company of 1845. A waterwheel raised 200-pound stamps that dropped one foot onto the rock containing the native copper in order to break the rock and free the metal. From Jackson (1845b, frontispiece).

to the story, had advanced such views while facing great opposition from his geological contemporaries.

Third, this claim, whatever its original source may have been, simply cannot be supported by, and is specifically contradicted by, an examination of Houghton's work and his frequently expressed opinions.

Fourth, as early as 1844 and 1845 Charles T. Jackson had clearly recognized both the fundamental difference between the occurrence of native copper in the Keweenaw and its occurrence in other mining districts of the world, and had publicly advocated its value for mining in published documents.

And fifth, professional geologists who knew the district well, such as Foster and Whitney, though recognizing clearly the presence of the native copper, did not at that time concede to it any unique or special status. As late as 1849, they were still unwilling to acknowledge Jackson's views as constituting a "discovery" in any meaningful sense.

Does this mean that Charles Jackson was indeed the discoverer of the uniqueness of the Keweenaw? He certainly believed so. And yet, the abundance of native copper had been the object of rumor and excitement for centuries, and the potential of the district for mining, at least in the abstract sense, had been recognized for an equally long time. Any case in favor of Jackson would have to be more narrowly defined, and center on his officially expressed opinion as a member of a scientific community within which his credentials were generally acknowledged. Perhaps equally important, any discovery or prediction cannot be made in a vacuum. To be meaningful it must be confirmable in some way; it must make some statement about the real world that can be either demonstrated or refuted. Here a major difficulty occurred for Jackson. The success in mining he so confidently predicted and foresaw for the Lake Superior Copper Company at the Eagle River location was simply not achieved.

Foster and Whitney, naturally enough, were not anxious to contribute to the elevation of Jackson's status. For them, whether the native metal was present in economically meaningful amounts was an empirical question, one that could be answered only by the long-term success of a profitable mine. Although mining efforts were beginning all up and down the district after 1845, the question of their ultimate success was still unanswered. The hard question of the actual nature and quantity of the copper was about to be resolved in the only place that it could be. The issue was being submitted to the ultimate arbitrator, the last court of appeals, to nature itself.

The Mining Efforts Begin

The Lake Superior Copper Company, for whom Jackson had done the analyses in 1844, was one of the very first companies to attempt mining on a systematic basis, and it did so at the Eagle River location. There Jackson had confidently declared the potential of the native copper, but he also specifically noted that he did *not* anticipate its discovery in large masses. Most of the copper was found filling small openings in amygdaloid traprock adjacent to a vein, and he believed mining success would be achieved by extracting the copper in that more finely disseminated form.

At that site, earlier mining operations had stopped over the winter, and many of the miners had gone home. Now, new miners were hired and the work was resumed as the company vigorously tried to develop the mine. Several shafts were put down, a stamp mill was built, and crushing machinery, "joggling tables," settling troughs, and various other pieces of apparatus were installed during the summer of 1845. Jackson himself directed much of the work.[19]

He made chemical analyses of the copper-bearing rock, which he now also called "ore." In his report to the trustees he stated that some samples reached values, for copper and silver combined, of over three thousand dollars per ton of rock, an amount so high that it could not possibly have been a representative sample. He also could not resist predicting "that there is no danger of exhausting the ore even should it give out at the depth of 100 feet, of which there is no probability. . . . If the ore runs out at a considerable depth, say 200 feet, it will be a matter of little importance to the present generation, though it might be to posterity."[20]

These optimistic claims added fuel to the "copper mania" of the eastern cities and helped divert much speculative capital to the district, but such figures proved to be wildly unrealistic. For the workers at the mine, those directly engaged in getting the mineral out of the ground, the lure of the traditional ores of copper compounds remained, based on their previous experiences in Europe. The working miners were always suspicious of this native copper, and especially its persistence as the mine deepened; this was just the reverse of what they had expected. Finding a "soft soapy black substance" at a depth of fifty feet, Gratiot was encouraged, and his Cornish miners thought that this was a good omen, for in Cornwall this material was frequently found to overlay the gray oxide ore.[21] However, the omen proved insignificant, the gray oxide did not appear, and profitability remained only a dream.

And so, Jackson's grand predictions notwithstanding, the location did not become a successful mining enterprise. Jackson later attributed the lack of success to inexperience and the difficulties of access to the location.[22] These were undoubtedly real problems, but the result was the same. The first serious, well-funded effort to mine the native copper of the Keweenaw ended without ever approaching the margin of profitability, and the question of the relative importance of the native copper and the other copper compound ores was still unresolved.

It is ironic that, at this locality, the native copper did not become profitable as Jackson had predicted it would. Foster and Whitney did not miss the opportunity to throw this back at him.[23] In Jackson's defense, it was a first effort with no precedent to fall back on, and he did qualify his optimistic predictions, even in the original report. As late as 1854 Whitney still had to concede that "Hardly a location on Lake Superior shows more abundant indications of copper,"[24] but even after management changes and several company reorganizations in future years, profitability was not achieved. Successful mining of native copper was still an undemonstrated possibility.

To complicate the situation even further, there was one more blind alley. For generations, virtually everyone who had passed along the coast of the Keweenaw peninsula had known of "the green rock." This vein contained a variety of minerals that included the green copper silicate called chrysocolla. It outcropped directly along the shore of Lake Superior, on the point that forms the east cape of Copper Harbor, very nearly at the northernmost point of the entire Upper Peninsula. It without doubt had originally formed from the alteration of either native copper or a copper sulfide mineral, but it was an example of a genuine copper compound ore, and it had helped to mold the thinking of both Schoolcraft and Houghton.

The location of this vein was included in one of three leases originally held by a Mr. Raymond, a speculator from Boston who represented a group of investors that did not have the money to develop the claim. The capital for such development eventually came through the intervention of John Hays, a druggist from Pittsburgh who had faithfully dispensed prescriptions until he read a newspaper article about the new copper boom in Michigan (fig. 21). He then promptly decided that a trip to the north country would be just the thing needed to revive his failing health. Appropriately enough, Hays had the foresight to obtain some financial backing from a wealthy hunting and fishing friend, just in case anything interesting happened to turned up.

Hays made the long journey to the Keweenaw in the summer of 1843, apparently keeping his eyes and ears open as rumors flew. Listening carefully to the possibilities available for someone backed by real money, he soon struck up an agreement with Raymond that made Hays and his friend partners in the three claims. Hays's reception, on his return to Pittsburgh, was enthusiastic, and a result was the formation of the Pittsburg[h] and Boston Mining Company ("a very heavy company" noted local expert "Dad" Brockway later).[25] It and

Fig. 21. John Hays, the Pittsburgh druggist who was the moving force behind the opening of the Cliff Mine. Courtesy Michigan Technological University Archives and Copper Country Historical Collections, Michigan Technological University.

the Lake Superior Company were the first two well-financed companies in the district, and the first two to begin serious mining efforts. In the spring of 1844 Hays, eight Pennsylvania coal miners, and a geologist returned and visited the claim at Copper Harbor. Their first efforts centered on the "green rock" vein, and some shaft digging was done, but little was found worth pursuing, for the green copper-bearing minerals were not abundant. That fall, however, a continuation of the vein was located farther inland, near the military compound at Fort Wilkins, but this time the mineral was in the form of a black oxide of copper. The troops at the fort, finding little demand for their services in protecting the miners from the Indians (or vice versa), had been digging on the grounds and had found many black copper oxide boulders "of great density, richness and beauty."[26] These were subsequently traced to their origin in a well-defined vein, which proved to be an extension of the vein that formed the green rock.

With much fanfare two shafts were put down and twenty-six tons of black oxide were raised, the first regular and productive mining effort in the region. According to Hays, the ore assayed about 80 percent copper, and it was authoritatively proclaimed by "Dad" Brockway to be the "richest and most extensive in the world."[27] A national newspaper, hearing of the find, confidently predicted a yield of $6 million per year.[28] In fact, as mining continued into 1845, at a depth of fifteen feet the ore abruptly ran out, a result that Jackson had foreseen when he called the vein an "accident." The mined product, sent to the Roxbury Smelting Works near Boston, yielded less than three thousand dollars in copper from an investment of over twenty-five thousand, and the location was shortly thereafter abandoned.[29] So ended the attempt to profitably mine the ores of copper compounds at one of the best-known and best-studied sites in the region. The Lake Superior Company had now failed trying to mine native copper, and the Pittsburg and Boston had failed trying to mine a combined copper oxide ore. The nature and value of any form of Keweenaw copper was still uncertain and undemonstrated.

Enthusiasm within the district was beginning to fade, and there was little to show for the many thousands of dollars disappearing into holes in the ground. If there was any copper in the Keweenaw in meaningful amounts, it was not at all obvious, and the mood was beginning to change to pessimism as the old haunting doubts about this pure copper were again gaining force. The situation was soon to change, however, and in an abrupt and dramatic manner. As part of its agreement with Raymond, the Pittsburg and Boston also had rights to two other lease locations farther to the west. Both were near Eagle River, not far from the site being worked by the Lake Superior Copper Company.

The Cliff Strikes It Rich

Exactly who was responsible for first realizing the potential of the best of those two tracts is difficult to determine and may never be known for certain. According to Hays, in November 1844 he and two others were going by boat along the coast of the Keweenaw when they were blown ashore by a squall. They then went overland by foot to their location near Eagle River where, exposed in the face of a high cliff, he found a vein that contained some native copper and looked promising.³⁰ Other accounts suggest that Raymond, from whom the Pittsburg and Boston acquired the leases, was already aware of the vein, and other claimants have also been suggested.³¹ In any case, Hays wanted to report to his partners in the East on the status of their venture, but he could not find anyone to carry the news. He finally hired two Indian guides and started out himself on the eighteenth of December. By foot, sleigh, stage, and rail, he made his way toward Pittsburgh. On his way he made a brief stop in Detroit, where, Hays said, he "met Dr. Houghton, the state geologist. He examined some specimens I had brought with me and was astonished as well as interested in the discoveries made."³²

It would be very interesting to know more about exactly what happened in this meeting, and exactly what Hays showed Houghton. The new location from which Hays had just come was soon to become the fabulous Cliff Mine, the first mine to successfully produce native copper exclusively. There appears, however, to be some uncertainty as to exactly which ores Hays brought with him. In one account, the author states that Hays showed Houghton ores from the vein on the new location, which, as subsequent developments showed, would therefore have contained native copper. He also suggests that it was the news of this find that led to the enthusiasm of the Pittsburgh people.³³ Others have suggested that it was the black copper oxide ore from the Copper Harbor vein that Hays was carrying back east, which at that time was still regarded very highly and was still being mined through the winter of 1844–45.³⁴

Houghton's inclination toward ores of copper compounds and the fact that Hays described him as "astonished and interested" would, I believe, suggest the latter case. Houghton would indeed have been very much interested in a rich copper oxide ore, particularly from near the "green rock" that Houghton had studied and where he had seen some black oxide on his earlier trips. It seems less likely that he would have been much impressed by seeing more lumps of native copper, particularly in the light of his own experience with much larger pieces than any that Hays might have carried with him for such a long distance. Further, the true nature of the native copper vein at the new location was not realized until developmental work began the next summer, and by the time the first really large masses were found Houghton was dead.

Hays then resumed his journey, arriving in Pittsburgh early in January 1845, where his fellow investors were "surprised and much elated" by his account of what had been happening.[35] In March, Hays again returned north to Copper Harbor with more supplies and men, but he found that despite the continuing efforts there over the winter, the vein of copper oxide simply was not continuing to produce. The ore pinched out downward, and mining activity there soon came to an end. In the summer of 1845 Hays therefore transferred his company's efforts to the location near Eagle River, the site of his findings of the previous fall. Here, fortunes were to change dramatically.

At the new location, a vein a few inches wide that contained some copper and silver was found at the top of a cliff, the Greenstone Ridge, that was over two hundred feet high. Halfway down the face of the cliff the vein widened to two feet, where it contained a variety of minerals, but the lowest part of the vein was buried in rubble. Jackson and Whitney, who was already acting as Jackson's assistant, visited the location that summer. They indicated that while there was no direct evidence at the surface that would lead to the suspicion of a rich deposit, the widening of the vein downward merited a closer look at the base of the cliff. Whitney later admitted that at the time he had no suspicion whatever of the discovery that was about to be made.[36] Furthermore, the vein was precisely the type that Houghton had earlier warned against, stating that any attempt to explore such veins in the hope of finding significant amounts of native copper would be "useless."

Nevertheless, during the winter some German miners were put to work clearing away the talus and debris at the cliff foot. They soon discovered a loose mass of native copper, which encouraged their further efforts. Exposing the lower cliff face, they discovered that the vein was displaced about twelve feet eastward. At that point an adit, or tunnel, was driven into the cliff. And then, sometime in the winter of 1845–46 and some seventy feet into the rock face of the cliff, the "great mass" was struck, a tremendous piece of native copper that dwarfed anything ever found before, vastly bigger, for example, than the fabled Ontonagon Boulder.[37]

This discovery was, without any doubt, one of the truly significant moments in the long history of the Keweenaw copper district, and it came at a very opportune time. The rush had been losing its momentum, and the lack of success of even the well-funded and well-organized companies had been taking its toll. This new discovery amazed and astounded the mining community; it was, at last, an evidence that the copper was really there after all. Significantly, this was the first large copper mass ever to be found within the rock itself, as opposed to lying detached on or near the surface of the land. It therefore demonstrated, once and for all, that the large scattered erratics that had so long been found, such as the Ontonagon Boulder, had had their ultimate origin within the traprock of the district itself, and had not been transported from distant locations such as Isle Royale, as some had previously maintained.[38] As the mining effort continued to press on, masses of pure copper soon began to pour regularly out of the Cliff. Some weighed as much as *fifty* tons each.

Even Jackson was genuinely astounded. He had earlier recognized the possibility of mining native copper, but he also believed that large masses would be much too rare to be meaningful economically. Now he gushed about the "truly wonderful" mines, the "noble lumps" of copper, and he reminded everyone of his earlier opinions, suggesting that "those who were surprised that I recommended working mines for *native* copper, should come and see and they would believe [his emphasis]."[39]

Even with the amazing masses, however, profitability did not come easily. In July 1846 alone, for example, the Cliff produced over a half-million pounds of copper, compared with some five thousand pounds for the only other producer, the unprofitable location Jackson had analyzed for the Lake Superior Company. Yet the balance sheet remained unfavorable. The company had taken in less than nine thousand dollars on the sale of copper and had spent over sixty-five thousand. Not until stockholder resistance to additional expenditures was overcome and further assessments made did the mine, rich though it was, finally officially became a money-maker. The directors declared that a dividend of sixty thousand dollars on the six thousand company shares would be paid on May 21, 1849, the first such dividend ever paid by any mine on Lake Superior. Significantly, it was paid entirely on the basis of native copper as the only ore product (fig. 22).

Spectacular though the results from the Cliff were, the implications of its huge native copper masses were not clear for the remainder of the district. The possibility that ores of copper compounds might still prove profitable at other localities remained real in many minds. At Lac la Belle, on the southeast side of the peninsula, for example, some small veins

Fig. 22. The Cliff Mine location, as shown in the Foster and Whitney report of 1850. The tripodlike structure just right of center stands over a vertical mineshaft, up which the ore was hauled by cable wound around the drumlike whims, which were turned by horses.

rich in the sulfides of copper were found in rocks distinctly different from the trap that formed most of the mineral range. Houghton had earlier commented favorably on the mining possibilities of the ores containing compounds of copper from this general region, and the veins became the focus of a substantial mining effort.

The physical attributes of the Lac la Belle site were ideal, and the lure of the possibility of a mine based on the traditional ores was still strong. Foster and Whitney wrote that "no place on Lake Superior affords greater facilities for mining; and the efforts of the company deserve to be crowned with success."[40] Nevertheless, it was becoming increasingly clear that such ores were simply not going to be found in substantial quantities. The effort there, although backed by a considerable expenditure of money and substantial work at the site, simply could not make a profit. Meanwhile, at the opposite end of the range to the southwest, the phenomenon of the

Cliff was about to be repeated, and on what was in some ways an even more spectacular scale.

The Mighty Minesota

Since the removal of the Ontonagon Boulder in 1843, the southwestern end of the district had not shown much promise for mining. A few half-hearted efforts had been made, but by 1847 the Ontonagon area had become virtually abandoned. However, in the winter of 1847–48 Samuel O. Knapp, along with a small crew of helpers, was wandering across the region.[41] Knapp was another easterner who had headed west to take his chances in the district. His geological experience apparently consisted of having spent some years back East in a woolen mill and two years serving drinks in a tavern, but that lack was no hindrance; "he was simply the smartest operator that appears on the scene of early copper mining." He was looking for copper prospects near the Ontonagon River, not far from the original site of the fabled boulder. Knapp had been employed by a group of New York investors that held several claims, but he had found that none of them looked particularly promising. He happened to pass over lease location number ninety-eight ("good old ninety-eight," as it later was known) and was immediately intrigued. A curious line of shallow depressions ran along the southern slope of a hill, visible even under three feet of snow.

Subsequent closer investigation soon showed that the depressions were clearly artificial, and that they were following the course of a vein as it ran across the ground. After several hibernating porcupines were dispossessed, further digging into one of these depressions revealed a very large quantity of ancient, grooved hammerstones. Sam Knapp obviously was not the first person to find the location of interest. The depressions were the pits of ancient Indian miners who, apparently many centuries earlier, had used the hammerstones to beat the rock and free the copper. The vein they were following ran parallel to the rock structure of the area rather than across it, as had always been the case farther up the peninsula. After clearing another depression of accumulated leaves and weathered debris, at a depth of some eighteen feet Knapp found a mass of pure copper ten feet long, three feet wide, and two feet thick, that weighed nearly six tons. It was resting on crude oak timbers that held it several feet above the vein from

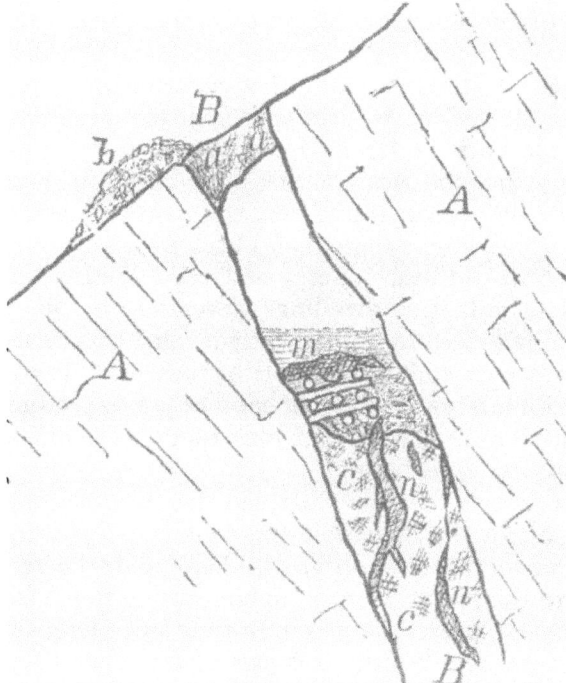

Fig. 23. Early sketch of a cross-section through the Minesota vein, drawn by Charles Whittlesey in 1856 and showing the vein's appearance at the time of its discovery by Sam Knapp in 1848. It had been explored to a depth of twenty-six feet by the ancient Indian miners. The six-ton mass supported by logs is indicated by "m," "b" is rock debris removed from the vein, and "n" indicates other large copper masses still in the vein. Note that the vein BB dips into the earth at an angle greater than the structure of the rock AA. From Whittlesey (1863), p. 17.

which it obviously had been detached and raised by the Indians at some time in the remote past (fig. 23). The timbers clearly showed the marks of small cutting tools, and the mass had been beaten and smoothed by the ancient miners' efforts to remove any projections for their use in making small tools and useful articles.

Equally interesting was Knapp's discovery that beneath the detached mass the copper continued downward into the vein below as a sheet of pure copper five feet thick. Knapp was quickly convinced that the location was very good, and on discreetly inquiring of another group of prospectors near the site he found that it was owned by a Detroit group of investors who had obtained it directly from the government. They might, he learned, be willing to part with it if the price was right. Back in New York he strongly recommended the purchase to his employers, and the deal was soon consummated. In the summer of 1848 a reorganized new company, the Minesota [sic] Mining Company,[42] sent Knapp back north again to begin work, which he did with "energy and zeal." Within one year he had built a mill, five dwellings, several other buildings, a two-and-a-half-mile road to the river, and had raised "500 bushels of potatoes and 300 bushels of turnips."[43] It was an auspicious beginning for a mine that soon began to produce more copper relative to the amount of rock removed than any mine on earth had ever done before.

As if to demonstrate that the Cliff had not been a fluke, masses of copper of equally incredible size soon began pouring out of the shafts of the Minesota (fig. 24). Here, with the vein running parallel to the rock structure, in contrast to those farther up the peninsula, shafts were soon strung out along the side of the hill. Foster and Whitney, as part of the federal survey, visited the Minesota location in the spring of 1849. Mining had not yet really begun in earnest, but they were still very much impressed. Even though the mine had been open for less than one year, they concluded that "no mine in the country has produced so great an amount of copper, with the same amount of labor and capital expended. We cannot find its parallel in the whole history of copper-mining, wherever prosecuted."[44]

Once again, profitability was not immediate. As always, the investment needed to begin was high, and it was 1852 before the dividends came, but once the mine hit its stride the Minesota became another wonder of the mining world (fig. 25). As mining continued, the metal proved so abundant that in sinking the second shaft it was feared that it would have to be moved to get around the huge masses of metal being encountered; digging a shaft through pure copper was not easy Soon there were over a thousand tons of copper visible and waiting to be chiseled into pieces and carried to the surface. The long-term impact that the mine had is perhaps best indicated by the ultimate accolade it received a half-century later when it was long past its glory days: "There is no question that

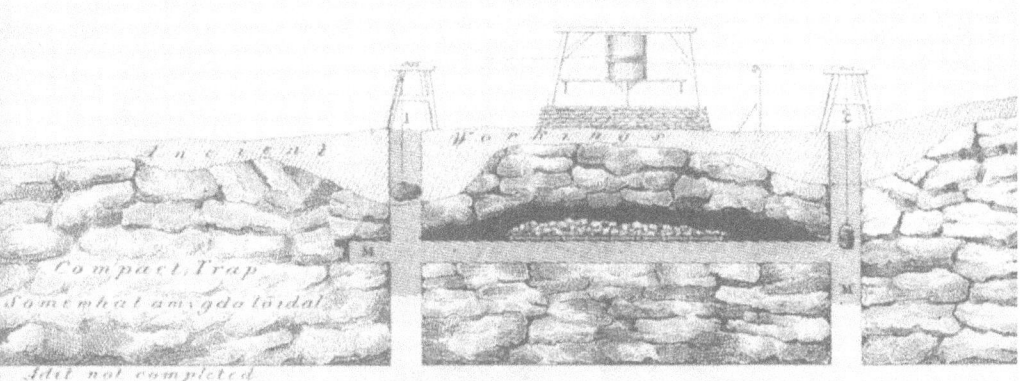

Fig. 24. Section of the Minesota Mine in 1849, as depicted in the Foster and Whitney report of 1850. Note that the vein had been excavated by ancient miners across the width of the view, and that shafts 1 and 2 were sunk into the deepest portions of those ancient workings. The copper shown hanging from the cable in shaft No. 1 marks the spot where the large mass was found that had been raised by the Indians and supported with logs (see fig. 22).

the Minesota deserves the title of the richest mine ever opened."[45]

The phenomenon was to be repeated once more, though on a somewhat smaller scale. A third mine, the Central, was opened on a vein a few miles up the peninsula from the Cliff. It also achieved great success in the mid-1850s by producing huge masses of native copper, so much so that it showed a profit in its first year of operation, "an unheard-of thing in copper mining."[46] These three mines, the Cliff, the Minesota, and the Central, became the talk of the mining world. Each produced many masses of native copper of immense size from the veins that cut the traprock, as well as large quantities of smaller masses called "barrel-work." They were testimonies to the boundlessness of nature, confirmation of the wildest dreams of every early prospector. They had an importance in the development of the district that far exceeded their simple production figures, and they became the standards by which every other claim was subsequently to be judged. Discovered at just the appropriate time, they reinvigorated the copper-mining effort. They also were an immensely spectacular evi-

Fig. 25. The Minesota Mine in the 1850s. An early picture of "the richest mine ever opened," taken about the time the huge 500-ton mass was found. Courtesy Ontonagon County Historical Society.

dence of one aspect of the true nature of the copper on Lake Superior.

Meanwhile, the possibility that the sulfide and oxide combined ores might still play a meaningful role in the production of the region, though always hovering everpresent in the background, was continuing to fade. After its great initial promise, the failure of the black oxide vein at Copper Harbor took its toll, especially in the reflected light of the unanticipated and sensational successes of the mass producers. The influence of the traditional ores lingered, however, in still providing the basic point of reference. Jackson, in praising the mass mines in 1849, wrote, "Truly they are *copper veins* [his emphasis]. The supply furnished by that mine [the Cliff] is as regular as it is in most mines furnishing ore."[47] Foster and Whitney, after their visit to the Minesota in the same year, similarly stated that "the veins of native copper are really valuable, and as certain in their yield as those of the *ores* of copper [their emphasis]."[48] As the two geologists continued the federal survey, they faithfully reported the minor veins of copper compound minerals that were still being pursued at various scattered locations, and they were still encouraging the Lac la Belle mine that was struggling to make the small sulfide veins there profitable.[49] It was, however, becoming increasingly clear what the result was going to be.

The Foster and Whitney Report

The appointment of Foster and Whitney to head the federal survey as the successors of Jackson had been "hailed throughout the mineral region with great satisfaction. . . . We have every assurance that under them, the work will be prosecuted with vigor and success."[50] They moved with dispatch to conclude the fieldwork during the season of 1849, for completion of the survey was vital to the sale of additional lands as the alternative to the now defunct lease system. Being in close contact with the companies attempting serious mining, they were also very aware that the wild days of "extravagant expectations" had largely passed, and "the business has settled down into a regular, methodical pursuit," with the promise of "a reasonable return for the capital invested."[51] Compiling their results over the winter, on April 15, 1850, they submitted *Part I. Copper Lands*, of their *Report on the Geology and Topography of a portion of the Lake Superior Land District* to Justin Butterfield, commissioner of the General Land Office. He submitted it to Interior Secretary Ewing, noting that it contained "a vast fund of valuable information, and the publication of it will be an important addition to the cause of science." It then went to Congress as an executive document and was ordered printed on May 16, 1850, with ten thousand extra copies ordered less than a month later.[52] This document, emerging exactly at midcentury, immediately became the definitive treatment of the copper district. It was, as late as the 1880s, still acknowledged as "having been for thirty years the most widely recognized authority on Lake Superior geology."[53]

In the report Foster and Whitney discussed in detail their contacts with the mining efforts in the district, noting in each case the nature of the product being pursued. Their final chapter on the mining aspects of the region was concerned with the treatment of the ores. Again they noted that the great majority of the copper ores of the world were sulfides, which required "a long and expensive process for smelting—a process which demands the highest metallurgic skill and a large amount of capital. The number of distinct processes which the ore undergoes in the great smelting establishments of Swansea is at least ten, which are entirely distinct from each other."[54]

And what about the situation in the Keweenaw? Here, at the end of their treatment of the district, they faced the long-standing issue squarely and head on: "The question arises then, are the sulphurets [sulfides] of copper destined to form an important part of the production of the Lake Superior mines?" Their entire final answer is worth quoting:

> The occurrence of the sulphurets of copper in the Lake Superior region is by no means a very uncommon fact. Two localities have furnished ground for mining operations—one of which has been entirely abandoned; the other is temporarily suspended. . . . We do not believe that appearances thus far indicate that any *ore* of copper exists in the Lake Superior region in sufficient quantity to be worthy of being worked. We have, therefore, a comparatively simple task to discuss the metallurgy proper of this district, since nature furnishes us with the pure metal [their emphasis]. [55]

Native copper, then, was not simply *a* copper mineral, or even *the* copper mineral, but the *only* copper mineral of the Michigan district present in amounts great enough to mine, a simple and direct conclusion that, after so much confusion and uncertainty, was subsequently confirmed and demonstrated by the totality of all the mining efforts in the district for the next hundred years. The question of the nature and value of the native copper of the Keweenaw was finally answered.

Or, somewhat more accurately, was *almost* answered.

9

The District Becomes Established

As the nineteenth century reached its midpoint, some major steps had been taken toward understanding and clarifying the nature of the Keweenaw district and its copper, and the speculative boom had given way to the serious business of mining. The copper-bearing ore minerals that almost everyone had expected to be there in such great abundance turned out to be present in amounts so small that they were of no economic value. Native copper, on the other hand and to the amazement of all, was soon pouring out of mines that had been opened at both ends of the district. Although much of the copper was in small "barrel-work" masses and even smaller "stamp-work" pieces, a surprising amount was in masses of very great size.

The Spectacular Fissures

The huge masses of pure copper emerging from the mines of the Keweenaw quickly became the hallmarks of the Lake Superior district, and some of them turned out to be so prodigious that they were almost unbelievable. Masses many tons in weight became common, and in 1856 the apparent fecundity of nature reached its zenith with a discovery made in the Minesota Mine. As the local paper reported it:

> The most astonishing developments have recently been made at the Minesota Mines. It would seem that wonders were never to cease on that location.
>
> If our friends below [in Lower Michigan] wish the description of a Lake Superior specimen, the existence of which we agree to vouch for, let them read the paragraph below, and they will find it to be something worth re-publishing.
>
> There is now in the Minesota Mine, between the adit and the 10 fathom level, a single detached mass of apparently pure metallic copper, which is some *forty-five feet* in length, and in the thickest part as much as eight or nine feet in thickness. If it is as pure as it appears to be, it contains probably more than *five hundred tons* of pure metal; and is certainly worth as it lies, more than *one hundred and fifty thousand dollars* [their emphasis].[1]

One million pounds of pure copper in a single "nugget!" No wonder the world was astounded. Who would have dreamed that mines could be found in which the major difficulty encountered would be simply getting all the copper out? Samples from these mines found their way into museums and private collections all over the globe. Geologists and mineralogists, whatever their former doubts and perhaps still against their better judgment, were soon convinced. Some went so far as to speculate that it might be possible to have too much of a good thing, that the cost of raising the copper to the surface might exceed its value. Attempts to split the masses with black powder explosions had little effect because the metal absorbed the force of the shock. Instead, they had to be slowly chiseled into more easily handled sizes that could be detached and sent up the shafts.

It is true that it would have been much more convenient if nature had provided the same amount of the metal in smaller, more manageable pieces. Nevertheless, the great masses did prove to be highly profitable under most market conditions, and their value generally far exceeded the cost of removal (fig. 26). This point was impressively made by the balance sheets and the dividend checks that were being sent to the smiling stockholders of the Cliff and Minesota Mines. By 1865 the Minesota had paid dividends of $1.76 million on an initial investment of $366,000, based almost entirely on the production of mass copper. The Cliff had paid an even more robust $2.1 million on an almost embarrassingly modest investment of only $110,000.[2]

Fig. 26. Miner and mass copper underground. An early Keweenaw miner poses proudly on a huge mass of native copper, the likes of which have been found nowhere else on earth. Courtesy Michigan Technological University Archives and Copper Country Historical Collections, Michigan Technological University.

The mining world marveled over what was happening, and among those who watched with interest was Henry Schoolcraft, who wanted to make sure that his role in emphasizing the importance of the native copper was not forgotten. In 1855 he published a *Summary Narrative* of his expeditions to the Northwest, in which he included, as an appendix, his original copper report to Calhoun of November 6, 1820. Comparison with the original, however, reveals an interesting change in wording that he made in the reprint.

In the original report Schoolcraft had written, in response to the reading he had done, that "no body of it [native copper] which is sufficiently extensive to become the object of profitable mining operations, is to be found at any particular point." Now, in the reprint of 1855, he simply changed the phrase "no body *is to be* found" to "no body *has yet been* found [my emphasis]." He thereby neatly turned his original disclaimer into a prediction that, he could now claim, was being fully confirmed by the success of the new mines![3] Not totally honest, to be sure, but indicative nevertheless of his actual original expectations and what mining was now revealing to be the case. After so much uncertainty, it now seemed clear that the age-old rumors and stories were true, unbelievable though they may have sounded before. The copper was really there in huge pure chunks just waiting to be taken, and the true character of Keweenaw native copper was now clear to all. Or was it?

In fact, despite the opening of these spectacular mines, despite the combined results of all the investigations of the preceding years, despite all the efforts of Schoolcraft, Houghton, and Jackson and all the detailed studies of the federal survey, the keys to the long-term success of the Keweenaw as a mining district still lay in the ground undiscovered. The Cliff, Minesota, and Central mines, in which the abundance of nature was so dramatically demonstrated by their gigantic masses of native copper, turned out to be quirks of nature. They were unique, and aside from these three, and despite a tremendous amount of searching, there was never to be another quite like them.

An examination of the records of the other mining efforts in progress throughout the district, and there were many, would have shown a far less encouraging picture. Ninety-one other companies tried mining before 1865. Despite the $12.5 million dollars that hopeful investors had paid in, the cumulative *total* dividends for all ninety-one companies came to less than the amount paid out by either the Cliff or the Minesota individually. Eighty-six of the companies had paid *no* return on investment whatsoever.[4] These were not very good odds of success, even for a risky business like mining.

The basic reason, we now realize, was that all the efforts in the district to this point had centered on one lode form only: the fissure, or vein, lodes. Once again, the mystique of the Houghton tradition has continued to obscure the course of events. In a recent discussion of Houghton's accomplishments, already examined in another context in chapter six, the author wrote that Houghton

> determine[d] the exact location of the copper deposits . . . and the nature of the different copper occurrences within the district. . . . His examination showed that the native copper always occurred in any of three situations within the Keweenawan basaltic lava flow sequence: in fissures within the flows, in the amygdaloidal zone at the top of a lava flow, or in the interflow conglomerate intercalated between two lava flows.[5]

Here, the discovery of all three of the lodes found in the district, the fissures, the amygdaloids, and the conglomerates, is clearly attributed to Houghton. Once again, however, this is simply not the case. A careful reading of Houghton reveals that all his remarks center on one lode type only, what he called the "true veins" and which he considered "simple fissures filled from below."[6] In the copper report he made this equally clear. There, he defined a "true vein" as "the mineral contents of a vertical or inclined fissure, nearly straight, and of indefinite length and depth." He added that since "the metalliferous veins being [are] contained under the head of true veins, it is to these that the whole of my remarks will be directed."[7]

Houghton was not alone in this conclusion. Virtually all investigations by all workers in the district before 1850 had centered on the possibility of copper ores in one lode form only: discrete veins that bore no direct relation to the regional structure of the rocks in which they were found. In short, these were what would now be called the fissure lodes, although it was not anticipated that the mineral of importance in them would be native copper. The mass mines that were opened on those lodes were essential to awakening interest and investment in the district. Without their dramatic impact the full development of the region might well have been delayed for many years. Yet they were, in a geological sense, "off the main line." Their role was quickly played out, and they were soon to pass from the scene as significant producers. The combined tonnage of copper produced from all the mines opened on these fissures, including the total production of both the Cliff and the Minesota, adds up to only about 2 percent of all the copper that eventually came out of the ground in the district.[8]

If the "true vein," or fissure lodes, did represent what the Keweenaw was all about, the district would have been just a "flash in the pan." After enjoying its few moments of glory, it would have soon earned the "I told you so" that so many had been ready to pin on it for so long. In reality, the amygdaloid and conglomerate lodes on which the long-term survival of the district depended were at midcentury still unknown and untouched. Certain clues to their existence were gradually being found, but their discovery and rise to profitability was to be a long and drawn-out process.

The early development of the one hundred-mile length of the district was by no means uniform. It proceeded roughly in thirds, beginning with Keweenaw Point to the northeast, then the Ontonagon region to the southwest, and finally the Portage Lake district in the center. These three regions correspond very nearly, respectively, to the present three counties of Keweenaw, Ontonagon, and Houghton (see fig. 18). As was discussed earlier, when the rocks of the district were forming, fractures developed that provided openings along which the copper-bearing fluids moved, and in which they deposited the native copper in masses ranging up to very large sizes. Up to midcentury and even beyond, these were generally considered *the* essential component of the district. Distinct veins of ore were, after all, the basis of the mining efforts in many of the classic districts of the world, and particularly at Cornwall, which had so often been used as the basis of comparison for the Keweenaw.

The Cliff Mine, the first and most sensational of the early developments, was in the northeastern region, where the copper lodes were of the classic form of crosscutting veins. This section produced most of the copper mined in the district well into the 1850s, with the production of the Cliff supplemented by a few less profitable properties opened on similar veins. In their 1850 report, Foster and Whitney declared that on Keweenaw Point there was only one system of veins. It had a bearing of about twenty degrees west of north, and was so clearly defined that "no permanently productive vein has been discovered thus far which varied 15° from this course, which is at nearly right angles to the formation."[9] In 1854 Whitney published his important book, *The Metallic Wealth of the United States*, in which he extensively reviewed the mineral resources of this country and compared them in detail with those of the rest of the world. He did recognize that there were other districts in the world in which copper ores were found in a second lode form, in which "the cupriferous ores are largely distributed through certain strata," but he clearly included Lake Superior copper in his former group, in which the copper was found "in true veins, in the trappean rocks."[10]

These fissure, or vein, lodes were nevertheless, despite their initial importance and promise, misleading for the future development of the district. They seemed to imply that in looking for copper prospects, attention should be directed toward searching *across*, nearly perpendicular to, the rock structure. However, the yet undiscovered amygdaloid and conglomerate lodes, which contained vastly more copper than the fissures, ran *parallel* to the rock structure at the surface (see fig. 6). The fissure lodes were also notoriously erratic. They could twist and turn, and a huge chunk of copper could be followed by great distances of barren rock, so that long-term planning was uncertain. From the point of view of steady production, although large masses were always welcome, smaller particles of copper distributed more uniformly through the rock would have made for a much more stable and predictable operation.

At the same time, it was becoming clear that this neat picture of crosscutting, perpendicular veins did not apply everywhere. At the southwestern end of the district near the Ontonagon River, for example, there were also fissure veins, but they were oriented differently. At the Minesota Mine the vein did not run across the rock structure, as the veins did farther up on Keweenaw Point, but instead ran parallel to it. Because of this unusual orientation, experienced miners at first viewed the Minesota with suspicion, and questioned the likelihood that working it would be profitable.[11] It nevertheless did prove to be a very rich true fissure vein, which dipped into the ground at an angle greater than the rocks themselves and thereby cut across the rock structure as it descended into the earth (see fig. 23). Foster and Whitney again commented on its resemblance to similar veins worked in Cornwall. By 1856, the year the great five hundred-ton mass was found, the Minesota and a few less profitable properties had made the Ontonagon region the most productive part of the district, accounting for nearly 50 percent of its production. Keweenaw Point was now in second place with about 45 percent, while the central region around Portage Lake produced almost nothing.

The productive portions of all these veins, then, whether running across or parallel to the rock structure, were confined mostly to the dark traprocks. Where such veins entered other types of rock such as conglomerate, their copper content had often fallen off abruptly. Of what they had designated mineral land in their survey of the district in 1850, Foster and Whitney had written that "all of that portion underlain by sandstone and conglomerate has been excluded—experience having demonstrated that, although they contain traces of copper, no valuable lodes need be expected."[12]

And yet, hints were appearing that suggested that there might be much more to the Keweenaw than this. During the winter of 1853–54, an unusual feature was found at the Copper Falls location, one of the first sites worked in the northeastern section. There, where all the veins were thought to be of the crosscutting type, investigations had brought to light a feature that Whitney declared was "unlike anything yet noted on Point Keweenaw. This is the occurrence of a metalliferous bed included in the formation and parallel with it."[13] The trend of this bed was marked at the surface by ancient Indian mining pits, and Whitney felt that, even though "great caution should be observed in giving an opinion in regard to it," it was well worth pursuing.

In the same year he similarly noted another discovery at the opposite, southwestern end of the district. In the National Mine, near the Ontonagon River and on the location immediately to the west of the Minesota, a "vein" was found that did not cut the rock structure at all. Instead, it was parallel to it, not only as it ran across the surface but also as it went downward into the earth. Whitney wrote that "the position of this vein is remarkable, and, up to the time of its discovery, entirely unprecedented on Lake Superior. It bears all the marks of a true vein, so far as regularity is concerned, but lies between two dissimilar formations [conglomerate and trap]."[14] The simple scheme of crosscutting veins was gradually giving way to the greater complexities of nature. It was, however, in the long-neglected central portion of the district that the new and truly significant developments occurred that held the keys to the long-term success of the Keweenaw.

The Amygdaloids Produce

This middle third of the Keweenaw district had always had some great natural advantages over the regions to the northeast and southwest. Not the least of these was the magnificent access provided by Portage Lake, which nearly cut the peninsula in two and which had provided, for so many years, the portageway that went completely across the copper-bearing series of rocks. Here, the possibilities for transportation, processing of ore, and the disposal of waste rock were almost ideal. Yet, in the early exploration of the Keweenaw this central portion had been largely neglected. The receding glaciers of the Ice Age had left layers of sand, silt, and boulder clay covering much of the surface of the land. Often these were many feet thick, and they hid much of the bedrock from the probing eyes of the prospectors and geologists. There was, however, one additional major concern.

Finding native copper had never really been much of a problem anywhere in the Keweenaw, at least in the traprocks. It was everywhere, but most of the time the amounts were very small. The problem was finding it in quantities great enough to make it pay. Near Portage Lake the copper that had been found was almost always in very small particles well scattered through the rock. These contrasted very unfavorably with the great masses emerging in the more developed areas to the northeast and southwest. As a result, the central Portage Lake district played little role in the very early mining history of the Keweenaw, or in the pioneering studies of Schoolcraft, Houghton, and Jackson. A few half-hearted attempts at mining had yielded little that inspired optimism, and the miners were discouraged by the lack of fissure veins.[15] At midcentury, the area was largely regarded as a barren stretch lying between the two richer extremities, but beneath its soil lay the real treasure of the Keweenaw.

In their report of 1850, Foster and Whitney gave an account of most of the mining efforts that were in progress throughout the district, along with synopses of their geological settings. The situation at one of these, near the middle of the region, was of particular interest. On the top of the high bluff that rose above the northern shore of Portage Lake, the Quincy Mining Company had been trying to initiate mining for some time.[16] The company had been formed by a merger of two earlier companies that discovered, to the surprise of each, that they held leases for the same piece of property. The early efforts, consisting of unsystematic probings of the rock beneath the topsoil, had not produced much.

When Foster and Whitney visited the location, the prospects were still so uncertain that they could express no opinion on the possibility of its ultimate success. They did, nevertheless, call attention to one unusual feature of the site. Although still referring to the lode that had been found as a vein, they noted that it lay between an upper wall of compact trap and a lower wall that was amygdaloidal. The native copper that it contained was "diffused through the veinstone. . . . The vein bears north 43° east, and dips rapidly to the north, corresponding with the course of the formation—the only instance of the kind observed on Keweenaw Point."[17] It was, in fact, an amygdaloid lode, but there was nothing at the time to suggest that it was of any value, and the future of the location did not appear bright. In the light of the masses from the fissure mines, the tiny, scattered copper particles of the Portage Lake district were almost universally regarded "with incredulity, if not contempt."[18] Freeing the copper would have been very difficult, requiring the entire rock to be smashed into virtual sand.

At least one man had faith in the region (fig. 27). Ransom Shelden, originally from upper New York state, had settled for a time in Wisconsin where his farming ventures did not prove particularly successful. While there he married Theresa Douglas[s], a cousin of Douglass Houghton, but they soon continued their wandering. In 1846 he arrived in the Keweenaw with his wife, two small children, and no money, but here he stayed on to become "the father of the Portage Lake district." The visionary Shelden joined his brother-in-law, Columbus Christopher Douglass, the cousin of Houghton who had been the geologist's assistant on many of his travels. Shelden ran a store, bought land, and dabbled in mines, and he was convinced that the mining future of the area was boundless. Offered a position at the mines near Ontonagon, he refused, "giving as his reason the opinion that there was more copper in the Portage Lake district than in Ontonagon and Keweenaw counties combined."[19] Later, in 1885, Shelden was described as

> this man, with limited education and scanty knowledge of geology and mineralogy—having been a tin-peddler in New York State and a peddler of essences and nostrums in the West—this man, full of pioneer energy and courage, was an enthusiast. He be-

Fig. 27. Ransom Shelden, the father of the Portage
Lake region. From History of the Upper Peninsula of
Michigan. The Western Historical Co. (Chicago),
1883, p. 285.

believed in the value of the mineral resources of his beloved Portage. His waking thoughts, his midnight dreams, were all of this hidden wealth. He believed in it; he talked of nothing else. The burden of his song was ever: Portage, Portage! So it came to pass that when he visited the other districts, he was considered, what would be called in modern parlance, a crank. But this man never flinched, never bated one jot or tittle of his high faith in his own district.[20]

Together with Douglass he became deeply involved in many of the mining companies around Portage Lake that were trying to become established at the time, including the Quincy that became one of the giants of the district. Douglass, despite his previous geological experience, was apparently not as certain of the prospects in the region, but Shelden "succeeded in inoculating his partner with some of his own enthusiasm," and they came to control much prime mineral land as well as being leaders in all aspects of the life of the community around the portage. Immediate financial success was not forthcoming, however, as freeing the scattered copper from the surrounding rock was not easy. The Quincy changed hands several times, and surrounding localities had no more success. Beginning in 1854, and after nearly a decade of frustration and failure, most of the mines of the Portage Lake area shut down.[21] The potential, however, remained, and Shelden was convinced that the metal scattered through the rock would soon come into its own.

Although profitability was elusive, the nature of the lodes was becoming clearer. In Whitney's classic study of that year, though noting the tentativeness of the future prospects because of the "peculiar" mode of occurrence of the copper, he nevertheless sketched the emerging picture. There were no veins of the usual type, but the copper was disseminated "through certain metalliferous beds, which run with the formation, and differ very slightly in composition from the other trappean beds with which they are associated. . . . In some instances, the same bed has been distinctly traced for a mile or more, by a line of ancient excavations, and, wherever opened, found to contain copper disseminated through it."[22]

The "ancient excavations" were the pits that the prehistoric Indians had dug to reach the copper, and similar ones had previously guided Sam Knapp to the vein on which the Minesota was put down. Once Douglass realized their significance here, they helped him recognize and define several new copper-bearing amygdaloid lodes.[23] A lode had also been opened just south of Portage Lake on another line of pits that ran along the trend of the rock. There, at the Isle Royale Mine, an attempt was being made to work what Whitney called a "metalliferous bed. It appeared to lie in an amygdaloidal belt of rock," and it was, he said, "as even in its bearing and thickness as a belt of sandstone under the same circumstances would be." Even granting the disseminated nature of the small particles of copper, he saw possibilities for its future. "If the copper is found to hold in sinking on this bed, it will be capable of furnishing a very large amount of that metal, and we may soon expect to see it rivalling with the most productive mines of Lake Superior."[24]

At the Quincy location, mining resumed in 1856 with C. C. Douglass at the helm. Several lodes crossed the property, but the real promise of success came with the full opening of the Pewabic amygdaloid. The Quincy put down its first shaft into the Pewabic late that year. The opening of this lode dramatically reversed the fortunes of the company, and three years later six shafts stood in a row, each a few hundred feet from the next, marking the trend of the lode as it ran across the property.[25] Locating an amygdaloid, however, even a rich one, did not ensure automatic profitability. The copper typically made up less than 5 percent of the rock removed from the mine. Elaborate and expensive stamping equipment was required to smash the rock and free the particles of the metal, and barren areas were still encountered. Nevertheless, in contrast to the fissure lodes, production could be planned in advance with at least some degree of certainty, and the opening of the amygdaloids was to bring at least a degree of stability to the copper-mining industry of Michigan. Gradually, led by the Quincy, the amygdaloid lodes began producing and Shelden's beloved region around Portage Lake started to prosper.

Still, putting money into the copper mines of Michigan remained a precarious investment. In 1865, with the Civil War over and copper prices down, the picture was gloomy. Of three hundred mining companies organized since 1845, only three, 1 percent of the total, had paid regular dividends: the Cliff, the Minesota, and the Quincy.[26] During that twenty-year period, even though prospectors had crisscrossed the peninsula in every direction, no more profitable fissure lodes had been found. The disseminated copper of the amygdaloids was slowly beginning to pay, but the greatest single body of native copper ore in the district still lay undiscovered beneath the ground. That situation, however, was about to change.

Houghton had originally visualized the major part of the trap range as divisible into two broad units as it ran the entire length of the peninsula: an amygdaloidal part to the northwest and the more massive part to the southeast. As more and more mines were opened and the rocks of the district were more carefully explored, it became clear that there were many subunits within the trap, including several amygdaloids that contained metallic copper in potentially mineable quantities. Occasionally, however, rather thin layers of sedimentary rock had also been discovered that lay between the trap units. These were usually conglomerates, but they were generally ignored as they did not contain enough copper to be of any economic significance. One of these layers did, however, and its discovery was to change the future of the entire region.

The Conglomerate Bonanza

Apparently, in any self-respecting mining district, there is a tradition that requires the discovery of its richest lode to be attributable to some form of animal life. In other districts the animals have included gophers, coyotes, prairie dogs, and roosters. In the Keweenaw they were pigs. According to one version of the local tale, one Billy Royal ran a roadhouse some miles north of Portage Lake on the path toward the Cliff mine. A herd of his pigs became lost by stumbling into an ancient Indian mining pit. After their squealing drew attention to the pit, further investigations revealed that the underlying bedrock was a rich, copper-bearing conglomerate, unlike any ever found before. Ralph Pumpelly, a prominent geologist who worked in the district in the 1870s, found the story attractive enough to relate that "I like to believe that credit was fundamentally due to the pig."[27] In reality, the discovery had little to do with luck or pigs.

Edwin J. Hulbert was a nephew of Henry Schoolcraft who, in acting as a secretary to his famous uncle, learned much about the Michigan district (fig. 28).[28] He had come to the copper country as a young surveyor and civil engineer in 1852 and was acquainted with many of the important figures of the period. A military road had been planned to run the length of the district from Copper Harbor to Ontonagon, and it was while employed on that project as a surveyor that Hulbert got his first hints of what was to follow. His first discovery was made around 1858, while working in the undeveloped area halfway between the Cliff Mine to the north and the amygdaloid lodes that were being developed around Portage Lake to the south. There, he found an intriguing piece of reddish conglomerate rock lying on the surface of the ground that showed "a fine percentage of copper" as part of the material cementing the pebbles of the rock. According to one story, he exhibited his sample to the local Cornish miners, but having apparently seen the likes of it elsewhere, they were not impressed. "Pudding-stone be a brave and 'andsome h'ore," they said, "but there bain't enough of h'it in the 'ole world to keep one man h'in baccy."[29]

Fig. 28. Edwin J. Hulbert, discoverer of the Calumet Conglomerate, the richest native copper lode of them all. Courtesy Michigan Technological University Archives and Copper Country Historical Collections, Michigan Technological University.

Subsequently, however, he found more pieces like it in the same general area. Deep in the woods he came across an immense, angular block of the same rock lying on the surface, and its sharp-edged appearance told him that it could not have been transported very far from its bedrock source. Although the bedrock was completely buried by glacial deposits, by plotting his findings he discovered that they all fell along a line that followed the structural trend of the rock units. Now convinced that a rich conglomerate lode existed within the bedrock beneath the surface, he discreetly began buying up land along his projected trend. He got financial backing from some prominent Boston investors, but several delays ensued. It was not until the summer of 1864 that he set out to find the source of his conglomerate fragments.

Late that summer digging began at a point that Hulbert had chosen on the basis of his projections. By mid-September the workers had reached the bedrock and completely laid bare a thick conglomerate layer enclosed between two trap units, and in which the reddish pebbles were cemented by finer material that included a significant proportion of native copper. Clearly this was the source of the pieces he had found on the surface and was therefore the lode for which he had been searching. Subsequent investigations revealed the rich conglomerate to be as much as twenty feet thick as it ran for close to two miles across the lands over which Hulbert had gained control, and it had good showings of copper for most of its length.

Later, when the lode had proved its value, others anxious to exploit its riches traced the conglomerate horizon carefully for many miles in both directions. When it left the land that Hulbert had selected, both to the northeast and to the southwest, it shrank to a thin seam of sandstone or shale that had no economic value whatever. By either incredible luck or amazing insight, he had gained effective control of virtually all of what came to be called the Calumet conglomerate, by far the most valuable lode ever discovered in the history of the copper district and the most valuable native copper lode on earth.[30]

Mining it was another matter. Two mining companies were formed, the Calumet and the Hecla, and Hulbert had substantial holdings in both. Unfortunately, with Hulbert in charge of the mining little progress was made. He tried to get profits quickly by digging huge open pits into the conglomerate, rather than using the shaft-mining approach which was clearly necessary for long-range success, because the lode dipped so steeply into the ground. The rich conglomerate rock, with its finely disseminated copper particles, proved to be much harder and more difficult to crush and process than the amygdaloid lodes that had been the basis of Hulbert's experience. A discoverer he may have been but a mine manager he was not, and his blunderings resulted in his soon being relieved of his position at the mines.

As a result of his failure to meet subsequent assessments on his extensive holdings in the properties, Hulbert lost all his shares in the Calumet and Hecla mines. He died in Rome, embittered and convinced to the end that he had been cheated out of his fair share of what came to be one of the richest mines on earth. However, his role in the discovery of the Calumet conglomerate, the greatest native copper lode of them all, is undisputed, even if he did not share in the vast wealth that it brought.

Meanwhile, as the 1860s went on and the amygdaloids and the great conglomerate were gradually coming into their own, the output of the spectacular fissure mines that had started it all was declining into insignificance. The year 1870 proved to be a watershed for the district in several ways. The glory days of the Cliff, the mine that had burst on the scene just when so much else seemed so discouraging, and the first mine on earth to produce its copper exclusively from native copper ore, were fading rapidly. Through the late 1860s its production dropped dramatically, and in June 1870 the *Mining Gazette* reported that the pumps had been stopped and the mine was filling with water; "the fate of the old mine is sealed."[31] At the opposite end of the district near the Ontonagon River, the mighty Minesota had likewise fallen on bleak times. In 1870, after seeing the dividends plummet, the company gave up on it. The mine was put out on tribute in which "buzzards," crews of miners that worked without company supervision, scavenged the mine in return for a share of what little copper was left. At both extremities of the district, the mines that had earlier pointed the way were now falling by the wayside.

At the same time, the finely disseminated ores of the central region were increasingly being mastered. Soon, the amygdaloids of the late-blooming Portage Lake region and the conglomerate of the Calumet and Hecla mines to the north, now under greatly improved Boston management that was willing to invest the needed capital, were increasingly dominating the production of the district. At the end of 1869 the Hecla paid its first dividend of $100,000, and in 1870 the Calumet did likewise. Because the two mines were working the same lode and had many shareholders in common, in early 1870 a consolidation was called for of the two companies. The logic of the union was unassailable, so in 1871 the Calumet and Hecla Mining Company was born, and it was to dominate the region for many years to come.[32] With its creation the district took a giant step toward maturity.

As for the nature of the copper of the Keweenaw and the lodes in which it was found, the era of uncertainty and speculation was largely over. By 1870 those questions were essentially answered. Other lodes were to be encountered in future years, but they would be of the types that were now well known. The nature of the district was firmly established, its course was set for many years to come, and the Keweenaw was ready to take its place as one of the greatest of the copper-producing districts of the world.

10

THE MAKING OF A MINING DISTRICT

Throughout much of the remainder of the nineteenth century the Michigan copper district dominated the copper mining industry of the United States. Between 1866 and 1884 its copper output multiplied fivefold as additional amygdaloid lodes were located both north and south of Portage Lake, and the production of the district continued on an upward curve well into the twentieth century. The former wilderness became the home of thousands of miners and their families as towns sprang up around the mines opening up and down the range. Many companies tried their luck at mining, but none more successfully than the Calumet and Hecla. Its "bonanza years" were beginning, and under very competent Boston management it became a wonder of the mining world. The metal flowed from the red conglomerate to provide the long series of regular dividends that enriched its stockholders and contributed so much to the flourishing of culture and philanthropy in New England.[1]

As the years passed, other companies also achieved varying degrees of success along the length of the district, but the profits were inevitably skewed strongly in favor of the most successful companies. The Quincy became another steady producer, drawing on its amygdaloidal lodes north of Portage Lake. For every real success story, however, many more companies were unable to even approach the uncertain line between profit and loss. For them, the money invested was "as completely lost as if it had been thrown into the sea," and they left little to show for their efforts but some crumbling buildings in a forest clearing.[2] Even so, hope springs eternal, and investors were always willing to bet that the next one would be another "C & H."

To many people, both inside and outside the region, the Calumet and Hecla Mining Company *was* the Keweenaw. Its huge profits more than countered the losses suffered by the dozens of companies that struggled on unsuccessfully, and were largely responsible for keeping the overall balance sheet of the district in the black. Calumet and Hecla was run efficiently, mined the richest lode, paid the best wages, and housed its employees adequately and inexpensively. Its management knew well how to look after the company's own best interests. Company workers found all aspects of their individual and community lives, even the most mundane, permeated by its benevolent paternalism: "The company was wonderful —a man always came and fixed the toilet."[3] Such attention to detail provided Calumet and Hecla with a stable, prosperous, and productive work force within which, not incidentally, radical political and social views would be unlikely to establish a foothold.[4]

The C & H copper emerged from seventeen shafts that plunged deeply into the earth and whose shafthouses stood at attention in a row along the two-mile length of the conglomerate outcrop (fig. 29). The depth of the mines, the gargantuan size of the mining equipment, and the distinctiveness of the native copper ores all combined to give the region a worldwide reputation. The small amount of silver naturally present in the copper enhanced its desirability for certain electric applications, and "Lake Copper" long sold for a premium price on the metal markets. It became the standard against which the copper produced from the more common sulfide ores of the other mining districts of the world was judged.

Throughout the third quarter of the nineteenth century Michigan's copper output was unchallenged, usually totaling over 80 percent and at times over 90 percent of the annual production of the United States (fig. 30). The Keweenaw continued to be a major producer and a potent factor in the copper industry for many more years. At the turn of this century the output of the Keweenaw was continuing its upward climb in tons of copper mined. Twenty thousand men were employed by the mining companies, and the district had a popu-

Fig. 29. General View of Calumet during the "bonanza years," looking northeast. The shaft houses are strung out in a line that follows the trend of the conglomerate lode and are surrounded by homes and businesses. Courtesy Michigan Historical Collections, Bentley Historical Library, The University of Michigan.

lation approaching a hundred thousand.[5] Annual production within the Michigan district reached its peak in 1916, when almost 270 million pounds of native copper came out of the mines in that single year.

However, a new factor had appeared on the national mining scene in the late nineteenth century. Vast deposits of minerals, mostly sulfides that contained small amounts of copper combined with other elements, were found in various localities in the western United States. In contrast to the native copper and the deep shaft mines of Michigan, these western properties produced vast tonnages of low-grade combined ores from huge pits that were open to the sky. They soon came to the forefront of the American copper mining industry and have remained there ever since. In 1888 Michigan yielded its national leadership in production, never to regain it again.

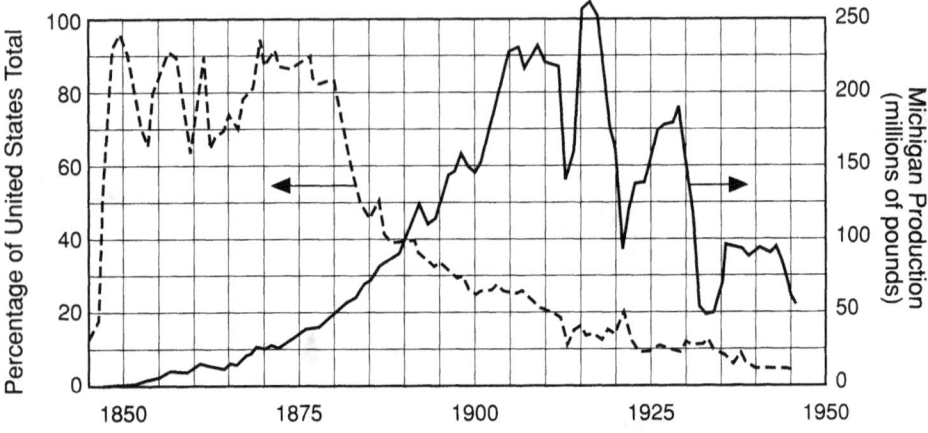

Fig. 30. Michigan and U. S. copper production, 1845–1946. The solid line is the Michigan annual production in millions of pounds, and the broken line is the Michigan percentage of the total U. S. production. Note that even as Michigan's copper production was increasing, its percentage of total U. S. production was falling, reflecting the opening of the low-grade western mines. Slightly modified, from Gates (1951), p. 195.

Over the years the Keweenaw was also subject to the usual ups and downs of the general economic life of the country. At first, each rebound pushed production to new heights, but every ton of ore taken from a mine is a ton no longer available for tomorrow. Inevitably the rising curve had to reverse its course, and it began an erratic but definite downward trend as the mines became ever deeper and the grade of the ores poorer. In 1920, on the bluff high above the north shore of Portage Lake, the Quincy Mining Company installed the largest steam hoist ever built to raise its ore from the lower levels of the mine. The shafts were approaching a depth of two miles down the slope of the amygdaloid lode, and the impressive nature of the costly machinery could not conceal the fact that the thin edge between profitability and loss was becoming more difficult to find.

The district never recovered from the effects of the Great Depression. At the Calumet and Hecla, which had since absorbed most of its smaller and less profitable neighboring companies through consolidation, a momentous and ominous moment was reached in 1939. Seventy-five years after it was first opened by Edwin Hulbert, the Calumet conglomerate, the "mother lode" on which the company had been founded, was abandoned and the fate was sealed of "one of the most famous underground copper mines the world has ever seen."[6]

The production of copper within the district continued for several more decades but, ironically, much of it did not come from mining. Over the years a disturbingly large percentage of all the mined copper had been lost, unavoidably it seemed. It had been washed into the inland lakes that were the repositories for the fine "stamp sand" left by the ore-crushing process. These losses had always been something of an embarrassment to the mining companies, but the introduction of new metallurgical and grinding techniques allowed these waste materials to be dredged from the lakes and processed again, thereby reclaiming much of the previously lost values. Throughout the first half of this century this secondary production helped greatly compensate for the increasing cost and decreasing grade of the ore that still came from the dwindling number of active shafts. In this way, by literally mining the same rock twice, both the Quincy and the Calumet and Hecla significantly prolonged their lives as producers of native copper.

The end could not be postponed indefinitely, however. Hopeful discoveries of new lodes did not compensate for the economic exhaustion of the old ones and the gradual depletion of the stamp sands. The Quincy, long nicknamed "Old Reliable" for its earlier half-century string of unbroken dividend payments, ended its operations with the closing of its reclamation plant in 1967. Shortly thereafter Calumet and Hecla lost its status as an independent company by being merged into the Universal Oil Products conglomerate and was also hit by prolonged labor strife. Finding itself unable to resolve the bitter dispute, Universal Oil Products "ignominiously and unnecessarily" shut down the last few operating shafts in 1969.[7] Now many of the towns are only shadows of their former selves, and the forests are reclaiming the clearings. Few remain of the spectacular mining structures that formerly dotted the landscape and so clearly indicated the economic basis of the life of the district. The Quincy shafthouse and its giant steam hoist, once the wonders of the mining world, now stand in silence on the north bluff of Portage Lake over an empty hole in the ground nearly two miles deep, lonely representatives of a ever receding and vanishing past (fig. 31).[8]

Despite occasional sporadic attempts to reopen various mines, the production of Keweenaw native copper has ceased. Examination of the figures for the life of the district reveals that nearly 40 percent of the vast total of the copper came from one single lode, the Calumet conglomerate. Five different amygdaloid lodes together contributed some 55 percent. The fissure lodes that started it all yielded only about 2 percent of the total, and the small remainder came from scattered properties.[9] All of it was native copper. An era that began with the discovery of the first large mass in the Cliff fissure in the winter of 1845–46, an era unique in the history of world copper-mining, has come to an end.

It has now been more than a century and a half since the beginning of the decade of the 1840s that initiated the events that made the Keweenaw one of the world's great mining districts. How has the claim for its uniqueness as a producer of native copper stood up through the years? Perhaps surprisingly, and despite the discoveries of new sources of copper ores throughout the world, that claim continues to remain essentially unchallenged. The district is still the only copper-mining region on earth where the primary ore mineral was native copper. Districts have been opened more recently, such as those in the western part of this country, that have produced vastly greater total tonnages of the metal, but they have done so by mining the traditional ores of copper compounds, particularly sulfides. However, two significant mining developments have taken place that might, at first glance, be construed as challenges to the uniqueness of the Keweenaw. One of these was in South America, and the other was within the borders of the district itself.

As has been repeatedly noted above, native copper is not a particularly rare mineral, but in most regions where it is found it had formed from the alteration of other copper minerals, such as the sulfides. Even in those areas where it does occur as a directly deposited mineral, it is normally not present in amounts great enough to serve as an ore. At least one apparent exception to this general rule, however, is found in the copper deposits at Coro Coro in Bolivia.[10]

There, a large series of copper deposits containing mostly ores of copper compounds is found extending for long distances across a high plateau. At one location in the region,

The Making of a Mining District 245

Fig. 31. The Quincy location in the 1960s. Broken windows and rusting sheathing of the Quincy No. 2 shaft house and its surrounding outbuildings recall the long-vanished glory days of Michigan native copper mining. Courtesy Fred Harley, Michigan Technological University Archives and Copper Country Historical Collections, Michigan Technological University.

Coro Coro, native copper is found in some abundance. From about 1830 to 1912 it was virtually the only ore mineral mined at that point, and in those years the mines there were often spoken of as analogs of those on Lake Superior. However, in more recent times it has become increasingly clear that the region abounds in sulfide ores. Even at Coro Coro, the sulfides are the copper-bearing minerals present in the largest quantities. These had been largely neglected in earlier years because of difficulties in processing them and the high cost of transportation from the remote location. The native ore is found only along a single fault zone, with a length of less than three miles, and it grades laterally into the usual sulfide ores. The comparison to the exclusively native ores of the Keweenaw is not, mineralogically speaking, a close one. It is true, then, that a few of the mines in Bolivia have produced native copper as their primary ore. At the same time, the distinction of the Keweenaw as a major district in which native copper is virtually the only ore mineral remains unchallenged by such developments beyond its own borders.

Equally interesting are more recent developments that have taken place within the Keweenaw district itself. Since the 1860s it has been known that at the extreme southwestern end of the mineral region, beyond the Ontonagon River and near the Porcupine Mountains, native copper was often encountered in a very finely disseminated form. There it was found, not in the conglomerates and lavas that carried the copper in the rest of the district, but in the younger, overlying sandstone and shale rocks. The Nonesuch location, just east of the Porcupine Mountains, was the focus of many of these earlier efforts, but attempts to work the ore were invariably unsuccessful, because much of the extremely fine copper was unavoidably lost during processing. Down through the years, recovery of the fine copper had been attempted "mechanically, metallurgically, electrolytically, chemically, courageously, and outrageously," but, unfortunately, always without the one essential: profitability.[11]

It had also long been known that there also was fine-grained sulfide mineralization in these sedimentary rocks. This sulfide ore may well have formed at the same time and by the same processes that produced the bulk of the native copper in the remainder of the district. However, in these younger rocks a fair amount of sulfur was apparently present in the sediments when they were originally laid down, perhaps

The Making of a Mining District   247

in the form of the mineral pyrite or even in the form of organic material. As the solutions that deposited the native copper in the rest of the district rose even further upward, the copper evidently reacted with the sulfur to produce the sulfide mineral chalcocite.[12] It too, however, was disseminated very finely through the rock, and there seemed little likelihood that its recovery could ever be made profitable.

Then, in the early 1950s, extensive testing indicated that new techniques might permit the chalcocite to be recovered and smelted successfully. Aided by a large government subsidy, the White Pine mine was begun to tap the chalcocite ore body. In 1955, for the first time in the history of the Keweenaw district, a copper mine was operated on an ongoing basis whose main mineral product was not native copper.[13]

And yet, this is an isolated case. The chalcocite ore body is distinct and separate from those that were mined earlier in the remainder of the district. The ore occurs in different rocks and in a different geological setting from the native copper ores. It played no role, nor would it have even if it had been known earlier, during the century of what might be called the classic mining period of the Keweenaw. This one sulfide ore body is to the remainder of the native copper of the Keweenaw, roughly, as the one native ore body at Coro Coro, Bolivia, is to the remainder of the sulfide ores in that district. The uniqueness of the Keweenaw again remains essentially unchallenged.

Finally, and for the last time, who *did* discover the uniqueness and value of the native copper of the Keweenaw?

A simple question with a complex answer. Recent historical research has clearly demonstrated that even the seemingly straightforward discovery of a simple scientific "fact" by one person can often disintegrate into a maze of qualifications, uncertainties, and even contradictions when examined closely. No less can be expected for a discovery that encompassed three and a half centuries and dozens or even hundreds of people. Nevertheless, a brief summary can be attempted.

At the first and most basic level, early native American Indians both knew of the existence of native copper in large amounts in the Lake Superior region and, at least at times, actively mined it in an organized and systematic way. Most such efforts, however, had apparently taken place long before the first European contacts with the New World.

At the next level, many Europeans, most of whom were not scientists or specialists in mining, became aware of the native copper deposits very soon after their first contacts with America. By the beginning of the nineteenth century the existence of native copper in the Keweenaw region and the possibility that it might well repay mining efforts was clearly understood, and some preliminary efforts at mining had already been tried. This "amateur," or public, tradition continued throughout the first half of that century and merged directly with the final demonstration of such success by the establishment of large-scale mining of the native copper around midcentury.

At yet another level, it is perhaps possible to speak of a more "official" discovery, in the form of a demonstration or argument by persons who were recognized as authorities by their peers and by society in general. They occupied positions that implied they had sufficient competence to be able to speak to the issue in an authoritative sense, rather than on the basis of uncertain reports, vague rumors, or dubious tales. The more basic question of the *meaning*, rather than the simple presence, of the native copper was the overriding issue. The emerging "professional" tradition, with its greater theoretical underpinnings and its greater sense of community, came to play a more prominent role here. Three persons were particularly important at this level.

Henry Rowe Schoolcraft. He lived at the beginning of the period that saw the growth of science as a profession in America, but his approach was still more representative of the amateur tradition. His introduction to the copper district came before he had become aware of what the presence of the native copper "should" mean, according to the theories of his day. Although his lack of formal education caused him concern, and he wavered as a result of his reading of the "experts," he never abandoned his original conviction that native copper would be essential in the mining future of the region. By the time mining had demonstrated the correctness of that conviction, his own interests had largely passed on to other things. Nevertheless, he clearly did play an important and pioneering role in the development of an accurate assessment of the value of the native copper.

Douglass Houghton. His is, without doubt, the most intriguing case of the three. Houghton's situation was the reverse of that of Schoolcraft, in that his education preceded, rather than followed, his own observations in the district, and this reversal may very well have been crucial. He, with total honesty, believed that his scientific methodology was truly empirical. There is little doubt that, had he lived, he would have been delighted to see the huge masses coming from the fissures and the native riches of the conglomerates and amygdaloids. That, however, was not to be. The past lay heavily upon his thinking, and he could never free himself from the weight of what he "knew" should be there.

Houghton's contributions to our knowledge of the district, and to the geology of Michigan in general, were unsurpassed by those of any other one person. On the question of the nature of the copper, however, he dramatically demonstrated the power of unacknowledged presuppositions. The copper district that emerged after his death was not the copper district that he had foreseen. Native copper, the attractions of which he feared would prove the ruin of so many, was in reality the only copper ore mineral present in the rocks that he knew. On the other hand, the minerals of copper compounds, for which he saw such great promise, were not present in those rocks in amounts great enough to profitably support even a single mine.

Charles T. Jackson. Although he was a part of the developing geological profession, he had two major advantages over both Schoolcraft and Houghton. The first was that his initial contacts with the district came later than theirs, and they were directly connected with potential mining efforts, as opposed to their earlier and more scientifically based interests. Profitability, rather than theoretical conformity, was for him a much more direct and immediate concern from the very beginning. Second, he had worked previously in Nova Scotia and New Jersey, where the geology was in many ways like that in the Keweenaw, right down to the rock types and presence of small amounts of native copper. His mental store of potential analogs, upon which he could draw for comparison as he worked in the Keweenaw, was the richest of the three, and he clearly was able to make the most of it.

It seems then, that Jackson does deserve more recognition than he has heretofore received for his work and insights in the Keweenaw. He clearly, by any standard, had serious personal and mental problems. Any attempt to honestly assess his scientific work is inevitably and understandably clouded by his involvement in the priority conflicts that have contributed so much to his present negative reputation. Even in the Keweenaw he did, to be sure, overstate his own case. At the location where he did most of his consulting work and where he believed this demonstration of commercial success would occur, it was not in fact attained. Nevertheless, even granting these qualifications, perhaps it is now time to recognize some merit in the priority claim that Jackson made for himself. Among those who can be identified as professionals in the study of the earth, Jackson was the first to clearly and publicly state his belief that in the Keweenaw, contrary to all the mining experience gained in all the other districts of the world, native copper was present in amounts that would be the basis for successful mining ventures.

And still, this was not the entire story. When mining success was finally achieved, Jackson was as astounded as anyone else by the immense chunks of metal that emerged from the great fissure mines that were the first to pay dividends. Even most significantly, none of the three, Schoolcraft, Houghton, or Jackson, had even the slightest suspicion of the existence of the amygdaloid and conglomerate lodes that subsequently proved to be the lifeblood of the district for the century that followed their discovery.

Examined closely and in some detail, then, the story is found to be a complex one with many strands. The geologists, as part of a newly emerging profession, were important and essential to the opening and development of the district. Their reports organized what was known. They made their information available to both governmental officials and private individuals. As professionals, their opinions carried increasingly great weight. Their views were quoted and requoted, and they drew the capital to the region that was so necessary for the establishment of successful, long-term mining. Without their influence, the development of the district might well have been delayed for many years.

And yet, perhaps strangely, it seems that their investigations, based on their expertise in the study of the earth, played little role in the emergence of an accurate picture of the nature and value of the copper of the district. In the nineteenth century it was often claimed that as science became a more professional discipline it would result in insights into the natural world that would go beyond those achieved by the simpler amateur approach that had dominated previously. To the degree that the opening of the Keweenaw can test that hypothesis, the judgment must be largely negative. The huge masses that emerged from the fissure mines to amaze and astound the world, and the later immense tonnages of copper that flowed from the disseminated amygdaloid and conglomerate lodes, were not being extracted from the ground by the intellectual heirs of the new professional geologists and mineralogists of that day.

With their painstaking surveys and carefully prepared reports, the professionals had vaguely pointed the way, but the uniqueness of the Keweenaw remained largely hidden from them. Instead, it revealed itself most clearly and directly to men such as John Hays of the Cliff, Sam Knapp of the Minesota, Ransom Shelden of the Portage Lake region, and Edwin Hulbert of the Calumet and Hecla. These men all stood outside the new professional tradition. They had come to the district as amateurs, with no theoretical axes to grind and no esteemed orthodoxy to defend.[14] They were the culmination of a long heritage that went all the way back to those first traders and explorers who saw pure copper in the hands of the natives and dreamed of finding its source. Such men were willing to take nature as it was, and if nature had a few surprises up its sleeve, well, so much the better. "Sulfides," "carbonates," and "oxides" meant little to them, nor did they regard the odds or the opinions of those in high places. Not impressed with fancy terms, they preferred to trust their instincts.

In their attempt to defend the priority of Houghton over that claimed by Jackson, Foster and Whitney wrote that "geology is eminently a science of observation." True enough, to be sure, but "observation" is a slippery word, even when it is invoked in the name of science. Ultimately, the nature of the Keweenaw and its native copper was not revealed to geologists and mineralogists, or even to the skilled practical miners of the day. To a surprising extent, those students of the earth were standing on the sidelines and watching. Instead, success finally came to those who were, perhaps to their advantage, largely ignorant of theoretical considerations and the outmoded lessons of the learned past. Motivated primarily by their own inner vision, a desire for riches, and an unbounded faith in the abundance of nature, they were willing to simply go out in the field, dig into the earth, and find out what was there.

NOTES

Introduction

1. In this book I will most frequently refer to the native residents of the North American continent encountered by Europeans as "Indians," as was done throughout the period being examined. For the spellings and contrasting interpretations of the word now rendered "Keweenaw," see Peters (1985), pp. 197–200, and Vogel (1986), pp. 134–36.
2. Young (1984), p. 9.
3. Lankton (1991). Lankton used the same quotation in introducing his study of the copper industry (p. vi).
4. Since the 1950s the White Pine mine in the Keweenaw has produced copper from an ore of copper sulfide. Its significance is considered in the final chapter.
5. Scholarly studies of the Lake Superior district include Taylor (1930), a Ph.D. dissertation that incorporates some analysis of the political background and governmental land policies that influenced the opening of the Keweenaw district; Hybels (1950), a 1948 master's degree thesis that was published as "The Lake Superior Copper Fever, 1841–47" and which remains perhaps the best account of the rush period; Gates (1951), *Michigan Copper and Boston Dollars*, an outgrowth of a Ph.D. dissertation, "An Economic History of the Michigan Copper Mining Industry, 1845–1946" which examines a century of the district's economic and financial background; and Dersch (1977), which expands on the social history of the district. Abbot (1971) and Hyde (1984) analyze the flow of investment capital to the district in its early years.

   Studies of individual mines include Benedict (1952), the story of the Calumet and Hecla mine that dominated the productive history of the district; Chaput (1971), the history of the Cliff, the first great mine of the region; and Lankton and Hyde (1982), a well-documented history of the

Quincy Mining Company. Thurner (1984), *Rebels on the Range*, analyzes labor relations in the mines, particularly at the time of the strike of 1913.

Popular works include Jamison (1939, 1950), which examine the Ontonagon area at the southern end of the mining district from the perspective of a long-time resident, and Murdoch (1943), a very readable but not clearly documented history of the Keweenaw called *Boom Copper: The Story of the First U. S. Mining Boom*. It has awakened the interest of many, including this author, in the history of the Keweenaw. Small popular studies of parts of the district continue to be produced, such as Clarke (1973–), a privately published series of booklets, *Copper Mines of Keweenaw*. Other local histories abound.

Historical accounts by geologists include Jackson (1849b), pp. 371-84, Foster and Whitney (1850), pp. 5-13, and Rickard (1932), pp. 222-30. Studies analyzing the efforts of Douglass Houghton include Allen (1915, 1922), and Martin (1945). George Merk (1981, 1982) has been particularly instrumental in calling renewed attention to Houghton's work in recent years. Hazen and Hazen (1985) comment on Houghton's efforts and include excerpts from his fourth annual report in their historical review of America's mineral industries before 1850. Their section on Lake Superior copper contains an unusual number of misstatements for a work whose intent is essentially bibliographic. Among these are: Schoolcraft did not "discover" the Ontonagon Boulder and it did not weigh two-thousand pounds (pp. 99-100); William Keating did not examine the copper region (p. 100); J. B. Griffin (1961) is cited as J. B. Bennett, and Calumet, Michigan, is cited as Calumet, Indiana (p. 104); and *Full Exposure* [1851] is misdated to 1849 (p. 105). On page 99 they state that Jonathan Carver's description of Lake Superior copper in his *Travels*, an account of his visits to the interior of North America in 1766 and 1767, may have encouraged the formation of a mining venture that took Alexander Henry to the district in 1771. The reverse is true; the mining venture influenced what Carver later wrote about the copper in his book, which was not published until 1778 (this episode is discussed briefly here in chapter 1).

1. Keweenaw Copper Before 1800

1. Quinn (1979), 1: 125. See also Brebner (1933), pp. 25, 93-94. On copper as the third of the precious metals, see Hazen and Hazen (1985), p. 91.
2. Ibid., pp. 285, 287.
3. Ibid., p. 274.
4. For comments on one aspect of eastern copper and the Indians, see Quinn (1974), p. 474. For the possibility that the copper was from the Keweenaw, see Kellogg (1925), pp. 344-45, and Griffin (1961), p. 32. Griffin is a valuable source of information on the question of the mining and use of Keweenaw copper by native Americans in the years before the European contacts with the New World. In the mid-nineteenth century Charles T. Jackson, familiar with the native copper deposits in New Jersey, believed it unlikely that the Indians would have located them and therefore he concluded that most of the copper found in Indian hands in the East probably came from Lake Superior. Jackson (1849b), p. 374.

5. Quinn (1979), 1: 307. According to Griffin (1961), p. 32, the word is close to a Mohawk word for metal.
6. Ibid., p. 327.
7. Ibid., p. 317.
8. See Griffin (1961), pp. 32–36, for a further discussion.
9. Costain (1954), chap. 5; Lanctot (1963), pp. 71–75; Eccles (1972), pp. 6–13.
10. Brebner (1933), p. 127.
11. Quinn (1979), 4: 407.
12. Ibid., pp. 329–31, 409–10. For comments on the minerals present in this region and their relation to the historical accounts, see Thwaites (1901), 3: 206.
13. Ibid., p. 434.
14. Griffin (1961), p. 147.
15. Kellogg (1925), pp. 59–60.
16. Wrong (1939), p. 135.
17. Wrong (1939), p. 242. It has been suggested that the reference here is not to Lake Superior but to copper found along the North Channel on northern Lake Huron. *Dictionary of Canadian Biography*, vol. 1, s.v. "Etienne Brulé," by Olga Jurgens.
18. Thwaites (1901), 38: 14, 237–39, 243.
19. Thwaites (1901), 45: 16.
20. Ibid., pp. 219–21.
21. Griffin (1961), p. 38.
22. Translated in Kellogg (1925), p. 347.
23. Griffin (1961), p. 44.
24. Thwaites (1901), 50: 265–66; 51: 65.
25. Thwaites (1901), 54: 153–65.
26. Karpinski (1931), p. 99: "No one can examine this fine delineation of Lake Superior and the northern parts of Michigan and Huron without amazement at the amount of scientific exploration and careful observation which made it possible. Not until the second quarter of the nineteenth century was any cartographical work of the magnitude and character of this Jesuit map executed in the Great Lakes area." The map was reproduced many times, including in a large format in the Foster and Whitney report on the copper district of 1850.
27. Thwaites (1901), 55: 99.
28. Kellogg (1917), p. 191; (1925), pp. 131, 348–49.
29. Kellogg (1925), pp. 349–50. Nute (1944), p. 160.
30. Kellogg (1925), p. 350. Kellogg's translation.
31. Many documents relating to this episode are found in Thwaites (1906), vol. 17. It is also outlined in some detail in Kellogg (1925), pp. 351–57, and, more briefly, in Griffin (1961), p. 44. Several writers ignore this French episode completely, and claim that the first attempt at commercial mining was that of the Englishman Alexander Henry some forty years later. Examples are Fuller, ed. (1939), 1: 559; Murdoch (1943), pp. 11–12; Chase (1945), p. 22; and Hybels (1950), p. 101. In reality, this earlier French attempt was more elaborate, had more governmental backing, and mined substantially more metal than the Henry effort, although it was not profitable in the long run.
32. Thwaites (1906), 17: 165–66.

Notes to Chapter 2

33. Ibid., p. 176.
34. Ibid., p. 234.
35. Ibid., pp. 237-40.
36. Ibid., p. 242.
37. Ibid., pp. 252-54.
38. Ibid., pp. 306-14.
39. The St. Anne being, apparently, the Iron River of Michigan, about thirteen miles west of the mouth of the Ontonagon River along the Superior shore.
40. Thwaites (1906), 17: 314-15.
41. Although official documents suggest that La Ronde was serious about his copper mining efforts, it has been suggested that he was really trying to get a free lease on the fur trade. See *Dictionary of Canadian Biography*, vol. 3, s.v. "Denys de La Ronde," by Bernard Pothier and Donald J. Horton.
42. Kalm (1937), pp. 524, 621. Also Fergusson (1981), pp. 64-65.
43. Thwaites (1906), 17: xviii.
44. By John Fiske, quoted in Russell (1939), p. 3.
45. Respectively, Chase (1945), p. 22, and Hybels (1950), p. 101.
46. Henry (1809), pp. 194, 204-5.
47. Nute (1944), p. 163; Chase (1945), pp. 22-23; Marshall (1887), pp. 336-40.
48. Carter (1979), 35: 4-12, gives an account of the history of the island. In later years the story clearly became garbled. On the maps compiled on the Houghton expedition of 1840, the present Manitou Island, just off the tip of the Keweenaw peninsula, is identified as the Island of the Yellow Sands. Peters (1985), p. 205, in discussing the origins of place names along the coast, has no explanation for this identification.
49. Henry (1809), chap. 7, provides the details of the venture.
50. *The American Magazine and Historical Chronicle* (1986) 2,1:2. (Publication of the Clements Library of the University of Michigan.) This issue deals extensively with Rogers and Carver.
51. Murdoch (1943), pp. 12-13, even gives detailed quotations from conversations between Franklin and several Europeans. Starbuck (1962), gives a complete discussion of the story. See also Griffin (1961), pp. 30-31.

2. Amateurs, Professionals, and Native Copper

1. Bornhorst, Rose, and Paces (1983) gives an excellent overview of the geology of the Keweenaw peninsula and its copper deposits, with detailed descriptions of many field localities and references to much recent literature. The most comprehensive treatment of the Keweenaw copper district from a detailed geological perspective that emphasizes the mining efforts is still Butler and Burbank (1929). Their extensive bibliography lists nearly five-hundred entries. Of more recent work, White (1968) is authoritative, also see White (1960). A complete analysis of the regional geological setting of the district is found in Wold and Hinze (1982). On a more elementary level, see Dorr and Eschman (1970), chap. 4.
2. It is difficult to give an exact definition of the word *lode* because, like many mining terms, it is used differently in different regions. As origi-

nally used by Cornish miners, lode meant an ore deposit very much like what I here call a "fissure lode," and that meaning would *not* have included the amygdaloids or the conglomerates. Ironically, local mining usage in the Keweenaw exactly reversed this original meaning: the amygdaloids and conglomerates *were* called lodes but the fissure deposits were not, and in their classic study of the district Butler and Burbank adopted this local usage even while recognizing that it was technically incorrect. Because complete consistency is apparently not possible, I will refer to all three forms, fissures, amygdaloids, and conglomerates, as lodes because they all are tabular deposits whose dimensions are great when compared with their thickness. All three differ from the massive low-grade sulfide deposits with no definite boundaries that constitute much of the copper ores of the western states and many other mining regions but which are clearly not lodes. See Butler and Burbank (1929), p. 63, n. 1 and p. 101, n. 4.

3. Pennebaker (1954), gives a useful, standard summary of copper minerals and ores.
4. See, for example, Bowen and Gunatilaka (1977).
5. Gates (1951), p. 7.
6. Dersch (1977), p. 300. See also Murdoch (1943), p. 17. Several writers give versions of the tale that officials of the British Museum were not even aware of the existence of native copper as a mineral until copper from the Keweenaw was brought to their attention in the early 1800s. See Williams (1907), p. 14; Murdoch (1943), p. 37; and Nute (1944), p. 156.
7. Ignoring native copper as it is found in the Keweenaw when classifying ore minerals continues, implicitly testifying to its rarity as an ore. Pennebaker (1954), pp. 21-23, for example, carefully (and properly) defines a copper ore so as to include the native mineral, but then ignores it as he states "Copper ores vary as regards type (whether sulfide or oxide)" and "Copper ores are commonly distinguished as 'oxidized ores' or 'sulfide ores'." The small amounts of native copper found in other districts are for convenience often included in the "oxide" ore category, but this classification is meaningless as applied to Lake Superior copper.
8. Butler and Burbank (1929) discuss in detail earlier theories for the genesis of the ore. Other work on the question is found in White (1968) and Stoiber and Davidson (1959). Robertson (1975) treats the minor sulfide deposits of the Mount Bohemia region. Livnat (1983) provides a complete analysis and interpretation of the deposition of the native copper with citation of much recent work.
9. Daniels (1967), p. 63. Daniels (1968) gives a complete account of this period.
10. Daniels (1968), p. 18. Originally called the *American Journal of Science and Arts*, the *"Arts"* was soon dropped from the title and the emphasis was always on science. References will be given as *American Journal of Science*.
11. Daniels (1967), p. 66, and (1968), pp. 36, 39.
12. Bruce (1987), p. 131. Bruce discusses the general question of the relationship between science and technology in nineteenth century America in his chapters 9-11.
13. Daniels (1968), chap. 8.

14. Greene (1984). In general, the professionalization and institutionalization of science has been viewed as essential and even necessary to the development of science in America. Little has been done, however, in looking closely at how the formation of such communities might act to inhibit or retard scientific progress. See Kohlstedt (1985), p. 17. Laudan (1977) has discussed one such instance. She found that the ultra-Baconian, antitheoretical geological philosophy held by the members of the Geological Society of London in the early 1800s strongly inhibited the scientific value of their efforts. However, for an alternative interpretation, see Miller (1986). Little also has been done to examine the views of "amateur" or "public" science that might have coexisted with, and perhaps even have been in competition with, the more professionalized groups that were developing. See Kohlstedt (1985), pp. 20, 24.

3. Henry R. Schoolcraft Meets the Keweenaw

1. Dunbar (1965), pp. 157-58.
2. U. S. Congress, *Attorneys General—Construction of Public Laws*. House Ex. Doc. 123, 26th Cong., 2d sess., p. 57.
3. Magnaghi (1980) provides a detailed account of this episode. However, Cass, in writing to Calhoun in 1819, stated that the agent was a Senator Tracy from Connecticut, and that he went as far as Michilimackinac. See Cass to Calhoun, Nov. 18, 1819, in Schoolcraft (1821a), pp. 303-4.
4. Schoolcraft (1821b), p. 210.
5. Carter (1943), 10: 328-30. Le Baron to Eustis, Sept. 30, 1810, received Nov. 9, 1810.
6. Ibid., pp. 328-29.
7. Ibid., pp. 336-37. Eustis to Coxe, Nov. 10, 1810.
8. Gallatin (1811), 1: 113-20.
9. Eustis (1818). Quotes in this paragraph are from this article.
10. Schoolcraft (1821a), p. 11. I have followed Williams, the editor, for the background of the expedition.
11. Ibid.
12. Cass to Calhoun, Nov. 18, 1819, in Schoolcraft (1821a), p. 302.
13. Ibid., pp. 303-4.
14. Calhoun to Cass, Jan 14, 1820, in Schoolcraft (1821a), p. 306.
15. Calhoun to Cass, Feb. 25, 1820, in Schoolcraft (1821a), p. 308.
16. Schoolcraft (1821a), p. 15. Osborn and Osborn (1942) provide the facts of Schoolcraft's life. For a less laudatory account, see Bremer (1982).
17. Osborn and Osborn (1942), p. 307.
18. Ibid., pp. 397-98.
19. Bremer, emphasizing the failures of Schoolcraft's early life, interprets his heading west to be "as much a flight from his disappointed family as a reconnaissance of new opportunities." Bremer (1982), p. 41.
20. Schoolcraft (1819).
21. Bremer (1982), p. 45.
22. Osborn and Osborn (1942), pp. 333, 457.
23. Schoolcraft (1851), p. 44.
24. Cass to Douglass, March 17, 1820, in Schoolcraft (1821a), p. 310.
25. Schoolcraft (1821a), p. 324.

26. Ibid., p. 107.
27. Ibid., p. 116.
28. Trowbridge journal, June 27, 1820, in Schoolcraft (1821a), p. 475.
29. Schoolcraft (1821a), p. 119.
30. Trowbridge journal, June 27, 1820, in Schoolcraft (1821a), p. 475.
31. Schoolcraft (1821a), p. 121.
32. Schoolcraft (1855), pp. 96, 98.
33. Schoolcraft (1821a), pp. 372–73.
34. Ibid., pp. 123, 125.
35. Schoolcraft (1855), p. 98. This recollection was, to be sure, written years after the events took place. The site of the boulder, the extreme southwest corner of section 31, T50N, R39W, is no longer accessible. It was partly flooded when the Victoria Mining Company built a small dam about 1900, and was completely submerged beneath the reservoir of the Victoria hydroelectric power dam built in 1930. Three sketches made on the expedition of 1820 (the frontispiece, fig. 11, and another in Schoolcraft [1855], p. 97) apparently provide us with our only visual representations of that location.
36. Trowbridge journal, June 29 and July 1, 1820, in Schoolcraft (1821a), pp. 476–77.
37. Cass to Calhoun, Oct. 21, 1820, in Carter (1943), 11: 66.
38. Schoolcraft to Calhoun, Nov. 6, 1820, "Report upon the Copper Mines." Calhoun Letters Received, War Department, National Archives, in Schoolcraft (1821a), pp. 344–52.
39. Journals of Douglass (p. 372) and Doty (p. 420), in Schoolcraft (1821a).
40. Schoolcraft (1821b).
41. Daniels (1967), p. 155.
42. Schoolcraft (1821a), p. 460. For the controversy, see pp. 21–22.
43. Daniels (1968).
44. "The Plough Boy," April 8, 1820, quoted in Bremer (1982), p. 45. Bremer finds Schoolcraft's strongly empirical approach to nature a partial explanation for the "disjointed character of most of his published works."
45. Schoolcraft to Nathaniel Carter, Oct. 27, 1820, in Schoolcraft (1821a), pp. 340–43.
46. Bremer (1982), p. 46.
47. Schoolcraft to Calhoun, Nov. 6, 1820, in Schoolcraft (1821a), p. 349.
48. Schoolcraft (1821b), pp. 201–16.
49. Cleaveland (1816), p. 450.
50. Schoolcraft (1821b), pp. 213–14. A further evidence of the reluctance with which Schoolcraft decided to go along with current thinking, and perhaps even of second thoughts he may have had over going as far as he did, may come from slight wording changes between the two versions of his report. In the November 1820 version, the quote given contains the phrase ". . . ordinary profits of mining, would be *materially* enhanced by occasional and disseminated *masses* of native metal" [my emphasis]. Notice the change of "materially" to "greatly" and of "masses" to "masses and veins" in the later version quoted above. This could be interpreted to mean that, despite the opinions of current authors, he nevertheless was trying to reassert his original views that were based on his observations.
51. In Carter (1943), 11: 298.
52. Schoolcraft (1822).

Notes to Chapter 4   259

53. Ibid., p. 10.
54. Ibid., pp. 10-13.
55. Ibid., pp. 13-14.
56. The vein can still be seen extending from the shore into Lake Superior. Access to the location is from the town of Copper Harbor by boat to Lighthouse Point.
57. Schoolcraft to Silliman, July 29, 1823, in Schoolcraft (1824), p. 44.
58. Schoolcraft to Calhoun, July 28, 1823, in Schoolcraft (1824), pp. 44-46. The ore was the copper silicate mineral chrysocolla.
59. A published account of these efforts of Schoolcraft in the Keweenaw will be found in Krause (1989b).
60. Carter (1943), 11: 332.
61. Cass to Benton, January 19, 1823, in Carter (1943), 11: 331-32.
62. Cass (1824), p. 3.
63. Ibid., p. 4.
64. Detroit *Gazette*, Feb. 28, 1823, reprinted in *Michigan Pioneer and Historical Collections* 7: 193.
65. Schoolcraft (1851), p. 242.
66. Ibid., p. 241.
67. Ibid., pp. 243-44.
68. McKenney (1827), p. 255.
69. Ibid., p. 458.
70. Ibid., pp. 463-64.
71. Ibid., pp. 473-74, and Schoolcraft (1851), p. 245.
72. Treaty between the United States of America and the Chippeway tribe of Indians; concluded Aug. 5, 1826, in McKenney (1827), p. 480.
73. McKenney (1827), p. 477.
74. Schoolcraft (1851), p. 245.
75. George F. Porter to Cass and McKenney, n.d., in McKenney (1827), p. 478.
76. For the general background of the expedition, see Mason's introduction, pp. ix-xxvi, of Schoolcraft (1834).
77. Schoolcraft to the Chippewa chiefs, Oct. 1, 1830, in Schoolcraft (1834), p. 110.

4. Douglass Houghton—Copper Finds Its Columbus

1. Russell and Hornstein (1899), p. 11, Murdoch (1943), chap. 3, and Joralemon (1973), p. 50. The latter phrase of the title for this chapter is taken from that of chapter three of Murdoch, in which he discusses Houghton. As will be seen below, it is perhaps a more appropriate title than even Murdoch himself realized. The major sources for Houghton's life and work are conveniently summarized in Wallin (1977), pp. 15-16 and appendix. See also the brief article on Houghton in *Dictionary of Scientific Biography*. Many details of his early life can be found in Bradish (1889). A useful short biography is Rintala (1954). Recent summaries of his life can be found in Merk (1982) and Milstein (1987). Almost all writings on Houghton, however, from the earliest to the most recent, verge on hagiography, and therefore must be read with caution, particularly regarding interpretations of his geological work. An interesting exception is the brief portrait of Houghton in the popular history of J. Martin (1944).

2. *Dictionary of Scientific Biography*, s.v. "Douglass Houghton," by Samuel Rezneck.
3. Information on the Rensselaer School is from Friedman (1981) and (1983).
4. Merrill (1924), p. 75.
5. Eaton (1831), pp. 151–52.
6. Houghton to Richard Houghton, April 25, 1829. Photostat in Houghton Papers.
7. Hubbard (1848), p. 218. Bradish (1881), p. 13, and (1889), pp. 13, 16, has somehow garbled several dates. He has Houghton graduating in 1828, even though he did not begin at Rensselaer until 1829. Another dating error will be noted ahead.
8. *American Journal of Science* 18: 200–201, 1830.
9. Houghton to Father Jacob Houghton, from canal boat *Surprise*, July 11, 1830. Photostat in Houghton Papers.
10. Eaton (1831), pp. 153–59.
11. Houghton to Father Jacob Houghton, Oct. 6, 1830, quoted in Friedman (1981), p. 20.
12. Eaton (1830), preface and p. 2. Houghton's signed copy of this book, presented to him by Eaton, is in the rare book room of the Hatcher Graduate Library of the University of Michigan, Ann Arbor.
13. Eaton (1830), p. 11.
14. Lyon to his sister, Nov. 24, 1830, in Lyon (1897), p. 444. See also Bradish (1889), pp. 13–14.
15. Bradish (1889), p. 17. Bradish, in true Horatio Alger style, suggests the "dime-in-the-pocket" detail. He also cannot avoid making Houghton "still in his teens" and "not yet twenty years old" when he arrived in Detroit. In fact, as Bradish certainly knew, Houghton had turned twenty-one that September.
16. Rintala (1954), pp. 12–13, quoting from Catlin Papers, Burton Historical Collection, Detroit Public Library.
17. Schoolcraft (1834,) p. xiv.
18. Schoolcraft (1855), p. 224.
19. Houghton to Richard H. Houghton, June 15, 1831. Photostat in Houghton Papers. Also in Schoolcraft (1834), p. 287–90, and in *Michigan History* 39: 474–80.
20. For an unsympathetic account of these changes, see Bremer (1982), pp. 57–58.
21. Schoolcraft (1851), pp. 355–56, 359–60.
22. *Detroit Journal and Michigan Advertiser*, Oct. 19, 1831, quoted in Rintala (1954), pp. 16–17.
23. Carter (1943), 12: 365–67.
24. In Schoolcraft (1834), pp. 290–94; also in Schoolcraft (1855), pp. 526–31.
25. *American Journal of Science* (1821b), 3: 210; (1824), 7: 45.
26. The phrase "the ores, ordinary in appearance but more important in situ" should be particularly noted. Its significance will be examined further as part of the question of the nature of the ores generally.
27. The situation is more complex than this account would imply. In fact, the traprock formation was not further upriver but was downstream, closer to the lake; the party had crossed it in the form of a range of hills on the way to the boulder, but did not recognize it as no bedrock was visible. The rocks of trap that Houghton noted in the stream may well have been

Notes to Chapter 4

carried further upstream (southeast) by glacial transport, and then subsequently eroded out into the riverbed above their original bedrock source. The essence of his analysis, however, remains sound. On his visit a year later he did see traprock in place before reaching the boulder.
28. U.S. Congress, House Exec. Doc 152, 22d Cong., 1st sess., pp. 17–20.
29. Schoolcraft (1834), pp. 290–94.
30. Houghton to Richard Houghton, Feb. 21, 1832, in Bradish (1889), pp. 117–18.
31. Houghton to John Torrey, March 20, 1832, in Schoolcraft (1834), pp. 294–95.
32. Houghton to Schoolcraft, April 3, 1832, in Schoolcraft (1834), pp. 295–96.
33. This expedition is well documented. Schoolcraft (1834), with Mason's editing, is a very complete source.
34. Houghton to Richard Houghton, June 24, 1832, in Schoolcraft (1834), pp. 297–98.
35. Torrey to Schoolcraft, Oct. 5, 1832, in Schoolcraft (1834), pp. 150–51.
36. His journal, as well as those of Houghton and Boutwell, a missionary, are included in Schoolcraft (1834).
37. Allen journal, June 19, 1832, in Schoolcraft (1834), pp. 180–81.
38. Ibid., Aug. 15, 1832, p. 230.
39. U.S. Congress, House Exec. Doc. 323, 23d Cong., 1st sess., pp. 7–68.
40. Schoolcraft (1834), Appendix F, pp. 352–69.
41. *Detroit Journal and Michigan Advertiser*, Oct. 8, 1834, quoted in Schoolcraft (1834), p. 366.
42. Houghton to John Torrey, Nov. 24, 1832, in Schoolcraft (1834), p. 305.
43. Houghton to Jacob Houghton, April 2, 1836, in Bradish (1889), p. 125.
44. For the boundary controversy, see Gilpin (1970), pp. 173–94, Soule (1897), and Stuart (1897).
45. It is therefore interesting to note that the occasional expressions of sentiment for separate statehood for Michigan's Upper Peninsula that are heard even today go all the way back to the prestatehood period.
46. Soule (1897), p. 380.
47. Schoolcraft (1851), p. 547.
48. Stuart, (1897), pp. 402–3.
49. Lyon to D. Goodwin, Feb. 4, 1836, in Lyon (1897), p. 475.
50. Lyon to A. Philes, Feb. 18, 1836, in Lyon (1897), p. 478.
51. Lyon to Andrew Mack, Feb. 21, 1836, in Lyon (1897), p. 479.
52. Lyon to Charles Hascall, Feb. 21, 1836, in Lyon (1897), p. 480.
53. Gilpin (1970), p. 187.
54. Lyon to John Barry, March 20, 1836, in Lyon (1897), p. 486.
55. Quoted in Dunbar (1965), p. 314.
56. Carter (1945), 12: 1177–79.
57. Sagendorph (1947), p. 228.
58. Detailed sources such as Soule (1897), Stuart (1897), and Gilpin (1970) do not mention Houghton in this context.
59. In a recent summary of Houghton's life, Merk (1982) states that Houghton was the major architect of the plan to include the Upper Peninsula within Michigan, an interpretation repeated by Milstein (1987). Merk writes (p. 39): "In an attempt to break the deadlock, the Michigan Territorial Council passed an act on December 26, 1833, calling for the appointment of three commissioners to negotiate a settlement to the

boundary dispute. Douglass Houghton was one of the commissioners appointed (Martin, 1945). Houghton proposed a boundary compromise in which the people of the Michigan Territory would relinquish to Ohio their claim to the Toledo Strip and in lieu receive the Upper Peninsula Chippewa lands and statehood." Merk repeatedly refers to this as "Houghton's proposal," to which, he states, after considerable delay, "the people of Michigan now agreed." Oddly enough, Merk's only reference for this interpretation is a one-page, totally undocumented summary of Houghton's life written in 1945 for a popular magazine (Martin, 1945), in which there is no mention of any commission, date, or specific proposal. The date of the authorization for the commission was December 26, 1834, not 1833, and it seems that no commissioners were appointed, for Indiana and Ohio rejected any possible compromise. Only a month later, on January 26, 1835, an act was passed that enabled Michigan to form a state government and a constitution, with the usual southern boundary that included the Toledo Strip. See *Laws of the Territory of Michigan* (1835), pp. 34–35, 72; Soule (1897), pp. 357–58; and Dunbar (1965), p. 309.
60. For general background on the state geological survey movement, see Hendrickson (1961) and Aldrich (1979).
61. Hendrickson (1961), p. 366.
62. *American Journal of Science* (1839), 36: 15.
63. Winchell (1886?). This is the most detailed account of the first survey of Michigan, written by Michigan's second state geologist, Alexander Winchell. For the dating of the document, see Winchell (1889), p. 133.
64. Allen (1922), p. 676. Also repeated in Rintala (1954), p. 32, and Milstein (1987), p. 10.
65. Dunbar (1965), pp. 249–50.
66. Winchell (1886?), p. 158.
67. Bradish (1889), p. 27.
68. *Detroit Daily Advertiser*, Jan. 10, 1838, quoted in Rintala (1954), p. 34.
69. *American Journal of Science* (1838), 34: 185–92.
70. Houghton to Stevens T. Mason, Dec. 1, 1837, in Houghton Papers. Also in Houghton (1928), pp. 83–89.
71. Houghton to House of Representatives, March 2, 1838, in Houghton (1828), pp. 149–52.
72. Houghton to Senate of Michigan, March 7, 1839, in Houghton (1928), pp. 340–45.
73. Houghton to Stevens T. Mason, Dec. 10, 1839, in Houghton Papers. Also in Houghton (1928), pp. 397–98.
74. Houghton to Stevens T. Mason, Dec. 1, 1837, in Houghton Papers. Also in Houghton (1928), p. 85.
75. Houghton to John Torrey, Nov. 1, 1839. Photostat in Houghton Papers.
76. Houghton (1928), p. 398.

5. The Copper Report and the Copper Rush

1. Winchell (1886?), p. 176.
2. Kohlstedt (1976) provides a detailed account of the early years of the American Association for the Advancement of Science and its predeces-

sor, the Association of American Geologists and Naturalists. The latter title was often shortened to the Association of American Geologists. See also Fisher (1981), p. 44.
3. Kohlstedt (1976), chap. 3.
4. Houghton to Woodbridge, Dec. 16, 1840, in Houghton (1928), pp. 480-81.
5. Bradish (1889), p. 57, insists that "to escape the possible imputation of taking advantages of his opportunities, Dr. Houghton avoided while in the employ of the state all speculations, though his chances for such were peculiar and abundant." This, however seems specifically contradicted by Hubbard's journals, and is one example of the care that must be exercised in reading Bradish on Houghton. More will be noted below. See Peters (1983), pp. 1-5, 12, n. 12.
6. Hubbard's journal, June 15, 1840, in Peters (1983), p. 47.
7. Penny's journal, June 19, 1840, in Carter and Rankin (1970), p. 32. This is the region within which were found the great iron deposits that were to become the second great mineral resource of the Upper Peninsula. The iron-bearing rocks do not outcrop along the lakeshore, however, but lie several miles inland, and were therefore not suspected at this time. Their discovery occurred just before Houghton's death.
8. Penny's journal, July 7, 1840, in Carter and Rankin (1970), p. 50.
9. Hubbard's journal, July 11, 1840, in Peters (1983), p. 65.
10. Hubbard (1874), pp. 56-58; also in Houghton (1928), pp. 61-63.
11. Penny's journal, July 11, 1840, in Carter and Rankin (1970), p. 54.
12. Houghton (1841), fourth annual report, in Houghton (1928), p. 558.
13. Penny's journal, in Carter and Rankin (1970), p. 55.
14. Hubbard's journal, July 20, 1840, in Peters (1983), pp. 75-79.
15. Penny's journal, July 20, 1840, in Carter and Rankin (1970), pp. 64-65.
16. Field notes, in Houghton Papers.
17. Houghton to Woodbridge, Dec. 16, 1840, in Houghton (1928), pp. 473-81.
18. Houghton to Woodbridge, Dec. 17, 1840, in Houghton Papers.
19. Houghton to Jacob Houghton, Jan. 17, 1841, in Bradish (1889), p. 127.
20. Houghton (1841), fourth annual report, in Houghton (1928), pp. 483-647. Houghton's part is pp. 483-569, of which pp. 528-59 concern the mineral veins.
21. Houghton to Woodbridge, Dec. 16, 1840, in Houghton (1928), pp. 473-81.
22. Ibid., pp. 476-77.
23. Houghton to Porter, Dec. 26, 1840, in Bradish (1889), pp. 113-16.
24. Lyon to Turner, Feb. 17, 1840, in Lyon (1897), pp. 531-32.
25. Hybels (1950) provides a complete, readable, and well-documented account of the copper rush period.
26. *Detroit Daily Advertiser*, May 20, 1841, quoted in Ehlers's manuscript notes, in Houghton Papers.
27. Hybels (1950), pp. 309-10.
28. Houghton to Sager, May 31, 1841, in Houghton Papers.
29. *American Journal of Science* (1841) 41:183-86.
30. Osborn (1942), p. 474.
31. Most of the details of the Eldred venture can be found in U.S. Congress, Senate Doc. 260, 28th Cong., 1st sess., April 1, 1844. An article by John

Jones, Jr., appeared in the New York *Weekly Herald* on October 28, 1843, that provided an extremely detailed, contemporary account of Eldred's efforts. The story is related well in Hybels (1950). Several early historians tried to attribute the removal of the boulder to local hero James K. Paul, one of the first potential miners to arrive in the region and the founder of the town of Ontonagon. This story has been repeated frequently, as recently as 1973 in Clarke (1973), No.1, *The United States Mineral Agency*, p. 12, and 1976 in Heinrich (1976), p. 96. The problem is that this story has only one source, Paul himself, who all agree was something of a "yarnspinner," and the documentary evidence for Eldred is overwhelming. It is perhaps possible that Paul was one of the unnamed men Eldred picked up on the spot to help in the effort. Moore (1895) and Jamison (1939), chap. 2 and p. 274, provide more details.

32. H. Messersmith to George Messersmith, June 24, 1843, in *Michigan Pioneer and Historical Collections* 7: 194.
33. Porter to Cunningham, July 24, 1843, in Fadner (1966), p. 165.
34. U.S. Congress, Senate Doc. 160, 29th Cong., 1st sess., Feb. 24, 1846, provides a detailed account of the permit-lease system along with some of its associated problems.
35. St. John (1846), p. 11.
36. Houghton (1842), fifth annual report, in Houghton (1928), p. 668.
37. American Journal of Science (1843) 45: 156, 332.
38. Osborn (1942), pp. 474–75.
39. Houghton to Jacob Houghton, Jan. 17, 1841, in Bradish (1889) pp. 127–28.
40. Rintala (1954), pp. 87–91.
41. Houghton to Jacob Houghton, March 13, 1842, in Bradish (1889), pp. 128–29. The date is incorrectly given in Bradish as 1841.
42. *American Journal of Science* (1844) 47: 115–16.
43. Houghton (1928), pp. 16–19.
44. Ibid., p. 681.
45. Brooks (1873), pp. 13–14.
46. Lyon to Johnson, Aug. 11, 1845, in Lyon (1897) p. 603.
47. *Detroit Democratic Free Press*, July 21, 1845, p. 2, quoted in Ehlers's manuscript notes, in Houghton Papers.
48. Quoted in Rintala (1954), p. 80.
49. Hybels (1954), p. 309.
50. On Houghton's death, see *Michigan Pioneer and Historical Collections* 22: 662–66 and Warner (1970), pp. 118–22, which includes a letter from Houghton's brother to his parents, dated Oct. 17, 1845, describing the circumstances of his death.
51. Thompson to Lyon, October 21, 1845, in Bradish (1889), pp. 90–91.
52. Bradish (1889). This work is the source most frequently used, often uncritically, by subsequent writers. Mott T. Green (1985), p. 98, has pointed out that much writing about geological personalities is characterized by the themes of "gratitude, reverence, progress, success, and concord. Criticism is minor, muted, oblique, and often may be entirely lacking. The author is almost always a friend, member, employee, or believer." This describes Bradish's effort perfectly, and it has contributed significantly to the skewed picture of Houghton generally accepted today.
53. Bradish (1889), p. 36.

54. Ibid., pp. 37-38. Rintala (1954), p. 58, also cites this episode as illustrating Houghton's courage and skill in the face of danger.
55. Hubbard's journal, June 12, 1840, in Peters (1983), p. 31. Thirty-four years later Hubbard (1874), pp. 21-62, recounted this journey, and recalled it as being "among the pleasantest of all my reminiscences of travel" but does not relate the episode noted by Bradish.
56. Houghton field notes, June 12, 1840, in Houghton Papers.
57. Penny's journal, June 12, 1840, in Carter and Rankin (1970), p. 22.
58. Bradish (1889), pp. 36, 39.
59. "Statement of facts connected with the drowning of Dr. Douglass Houghton . . . as related by Peter McFarland and John Baptiste Bodrie, October 14, 1845," in *Michigan Pioneer and Historical Collections* 22: 662-66.
60. Sherzer (1932).

6. Houghton—The Misunderstood Pioneer

1. Jackson (1846b), p. 150.
2. U.S. Congress, House, Report 591, 29th Cong., 1st sess., May 4, 1846, p. 3.
3. Winchell (1889), pp. 138-39.
4. St. John (1846), p. 34.
5. Faul (1983), p. 174.
6. Houghton (1928). In 1837, p. 86, he wrote that "it will be by the accuracy of the scientific portions that our survey, abroad, will stand or fall, and without this, the final report would be not unlike the block of marble, untouched by the sculptor." In 1838, p. 96, "those portions which may be considered of a strictly scientific character will be omitted until the final report may be made." In 1839, p. 167, "reserving the great mass of matter which has been accumulating, with the view to an elucidation of the condition and resources of our state, for a *final* report [his emphasis]." In 1840, p. 481, "The final and connected report which I am called upon to make at the close of my official labor must necessarily involve not only my individual reputation but also that of the State." In 1841, p. 483, "and since the whole will be embraced in a more perfect form hereafter." In 1842, p. 668, with perhaps a hint of uncertainty, "there still remains much to be done, in arranging the materials accumulated, for a final report upon the entire work." In 1843, p. 673, with a further glimmer of doubt, "I hope in due time to offer a final and systemized report, that shall embrace, in a condensed form, all that has been accomplished." In 1844, p. 679, "the greater amount of work has been performed in the office, in compiling and arranging the materials for the final report."
7. *American Journal of Science* (1845) 49: 82.
8. *Democratic Free Press*, June 11, 1845, p. 2.
9. Foster and Whitney (1850), p. 14.
10. Allen (1922), pp. 677-78; Allen (1915), p. 132.
11. Houghton to Jacob Houghton, Nov. 14, 1841, in Bradish (1889), p. 130.
12. Allen (1922), p. 681.
13. Houghton's annual reports are all found in Houghton (1928).
14. St. John (1846).
15. Ibid., p. 4.

16. Hubbard (1848). This item is unsigned. It would be interesting to know if the phrase "backwoods geologist" used here is derived from St. John, or whether they both derive from a single, original source.
17. Bradish (1889), p. 89, quoting from the Flint *Wolverine* of 1874. The article misdates Houghton's appointment as state geologist to 1838, and misdates his survey of the Upper Peninsula that led to the copper report of 1844.
18. Bradish (1889), pp. 33-34.
19. Ibid., pp. 38, 54.
20. Hubbard (1848), p. 221.
21. Houghton to Porter, Dec. 26, 1840, in Bradish (1889), pp. 113-16.
22. Hybels (1950), p. 105, 108.
23. Merk (1982), p. 41.
24. St. John (1846), pp. 48-49, says that Michigan sent Houghton to examine the country in 1835 (he accompanied Schoolcraft in 1831 and 1832), and that he made his official survey in 1843 (it was 1840).
25. Martin (1945), p. 10.
26. Bradish (1889), p. 53. This would have been the 1843 meeting.
27. Martin (1945), p. 10. Again, without documentation. Houghton did make his major presentation at the 1841 meeting at Philadelphia.
28. Bradish (1889), p. 54.
29. *American Journal of Science* (1840), 39: 190-91; (1841), 41:172, 183-86; (1843), 45: 136, 138, 155, 156,160, 161, 331-33, 348; (1844), 47: 104, 106, 115-16, 119, 127, 132.
30. Bradish (1889), p. 34.
31. *American Journal of Science* (1838), 34: 190-91.
32 Ibid., (1841), 41: 236, 242, 265, 267.
33. An examination of Hazen and Hazen's (1980) recent extensive bibliography of American geological literature revealed nothing, under Silliman or his son and coeditor, Benjamin Silliman, Jr., that, by title, might be so construed. A complete search was not made of all the publications themselves.
34. *American Journal of Science* (1842), 43: 249.
35. Schoolcraft (1851), p. 60, and Schoolcraft (1855), p. 364.
36. Butler and Burbank (1929), p. 11.
37. Schoolcraft (1851), pp. 242, 265.
38. Cleaveland (1822), p. 554.
39. *American Journal of Science* (1834), 27: 381.
40. Shepard (1832, 1835), p. 65.
41. Dana (1837), p. 395. See Kraus (1938), pp. 146-47, for comments on Shepard and Dana.
42. Examples are Winchell (1886?), pp. 160, 164-65, 183, 185 and Irving (1883), pp. 9-12.
43. Merk (1982), pp. 36-44. The review is Dott (1984), pp. 217-18.
44. Milstein (1987).
45. Merk (1982), p. 41.
46. Among the questionable claims for what Houghton discovered in this account are the following. The word *Keweenawan* as a geological term is meaningless when used with any reference to Houghton or his era. This term was not used in this way until the 1870s, some thirty years after Houghton's death. Another problem is that any conclusion about "400

Notes to Chapter 6   267

distinct basaltic lava flows" or "20 to 30 interbedded conglomerates" cannot grow out of the work of Houghton or his times. Such interpretations are of considerably more recent origin, and could not possibly have been inferred from the geological data available to Houghton in the early 1840s. On the introduction of the term *Keweenawan*, see Martin and Straight (1956), p. 152. A helpful summary of early interpretations on the nature and origin of the trap is found in Irving (1883), pp. 7-13, who contrasts Houghton's interpretations with those of later workers. He indicates that the interpretation of the trap as lava flows began in the 1860s. The most important of the interbedded conglomerates, the Calumet conglomerate, was not discovered until 1864.

47. Schoolcraft (1834), pp. 293-94.
48. Carter and Rankin (1970), p. 47. These editors have transcribed this entry in Penny's journal as "Our favorable symptom is. . ." but it would seems likely that the first word should be "One."
49. Houghton to Woodbridge, July 22, 1840. Photostat in Houghton Papers.
50. Houghton to Woodbridge, Dec. 16, 1840, in Houghton (1928), p. 477.
51. Houghton to Porter, Dec. 26, 1840, in Bradish (1889), pp. 113-16.
52. Houghton (1841), fourth annual report, in Houghton (1928), p. 539.
53. Ibid., p. 553.
54. Ibid., pp. 533-34.
55. Ibid., p. 540. Perhaps the full quote deserves some comment: "The excess of native copper (compared with the other ores) which occurs *in these portions* of the veins [his emphasis] is a peculiar feature, for it may be said, in truth, that other ores are of rare occurrence. In those portions of the veins traversing the trap, and where other ores do occur, it is usually under such circumstances as to favor the presumption that their origin is chiefly from that which was previously in a native form." Here, Houghton could be construed as putting forward the concept that the copper compounds formed from the native copper, rather than the reverse, which was the conventional view up to that time. However, his phrase "peculiar feature" and his use of the emphasis indicate that he is at this point speaking only of an uncommon portion of the veins, and is not putting forward a general account of the origin of the compounds from the native metal for the entire district.
56. Ibid., p. 554.
57. Rickard (1905), p. 36, and (1932), p. 229. In the latter, he wrote of Houghton that "curious to relate, he regarded the metallic condition of the copper as unfavorable to the persistence of the ore in depth , and he held this opinion until he ascertained that the contrary 'was more or less universal with respect to all the veins.'" A similar view is expressed by Heinrich (1976), p. 12: "Not inconsistently, he believed that the presence of copper in the native form constituted an argument against its persistence in the veins in depth. Later, after becoming familiar with the nature of the veins, he changed his belief."
58. Houghton (1841), fourth annual report, in Houghton (1928), p. 555.
59. Ibid., p. 558.
60. Ibid., pp. 556-57.
61. U.S. Congress, Senate Report 268, 28th Cong., 1st sess., April 3, 1844, p. 18.
62. Ibid., p. 23.

63. U.S. Congress, Senate Report 291, 28th Cong., 1st sess., April 9, 1844, p. 2.
64. U.S. Congress, Senate Report 98, 28th Cong., 2d sess., Feb. 11, 1845, p. 3.
65. U.S. Congress, Senate Report 175, Special sess., Mar 19, 1845.
66. Ibid., p. 21.
67. U.S. Congress, House Report 591, 29th Cong, 1st sess., May 4, 1846, pp. 6–38.
68. A few writers have briefly noted this component of Houghton's thinking. Alexander Winchell, who became Michigan's second state geologist in 1859, in a detailed commentary on the work of Houghton, recognized his conservatism on the ores in a single sentence when, in commenting on the copper report, he wrote, "His final prognostication is decidedly conservative, and it seems to be rendered so by the very circumstance which has constituted almost the exclusive resource for profitable copper mining in this district." See Winchell (1886?) p. 183. Murdoch (1943), p. 37, likewise comments in a single sentence, "Moreover, Douglass Houghton, an open-minded man if ever there was one, mentioned the native copper in his reports but far more frequently used the phrase 'ores of copper.'."
69. Bradish (1889), p. 54.
70. Beveridge (1950), p. 144.
71. A more detailed discussion of possible reasons why the "Houghton tradition" has lasted so long will be found in Krause (1989a).
72. Hubbard (1887), p. 78.
73. Quoted in Bradish (1889), pp. 111–112.

7. Charles T. Jackson and the Federal Survey

1. Peters (1983), p. 6. Winchell (1886?) provides a complete account of this period, and of the history of the first geological survey of Michigan.
2. Walker and Douglass to Lyon, Feb. 16, 1846, in Bradish (1889), p. 290.
3. J. Houghton and Bristol (1846); J. Houghton (1846).
4. Governor Alpheus Felch, Jan. 6, 1846, quoted by Winchell (1886?), pp. 198–99.
5. Houghton (1928), pp. 684–86.
6. Winchell (1886?), p. 201.
7. Hubbard to Bradish, July 18, 1850. Hubbard Papers.
8. Winchell (1886?), p. 201.
9. Talcott to Marcy, Jan. 15, 1846, in U.S. Congress, House Doc. 69, 29th Cong., 1st sess., Jan. 16, 1846, p. 2.
10. Tod and Bartlit to Marcy, Feb. 10, 1846, in U.S. Congress, Senate Doc. 160, 29th Cong., 1st sess., Feb. 24, 1846, pp. 20–25.
11. U.S. Congress, House Doc. 591, "Committee on Public Lands, Sale of Mineral Lands." 29th Cong., 1st sess., May 4, 1846, pp. 2–5.
12. In U.S. Congress, House Doc. 204, 29th Cong., 1st sess., May 20, 1846, pp. 1–2.
13. Ibid., pp. 8–18.
14. Hybels (1950), p. 323 and n. 256.
15. Foster and Whitney, pp. 15–16.
16. Brief biographic sketches of Jackson are found in *American Journal of Science* (1880), series 3, 20: 351–52; *Popular Science Monthly* (1881), 19:

404-7; *Proceedings of the American Academy of Arts and Science* (1881), 16: 430-32; and the *Dictionary of Scientific Biography* article by George Gifford, Jr. More extended is "Charles Thomas Jackson" by J. B. Woodworth (1897), which includes a lengthy bibliography.
17. Jackson (1849b), p. 401.
18. *First report on the Geology of the State of Maine*, by Charles T. Jackson, 1837. Augusta: Smith and Robinson, p. 86, quoted in Aldrich (1981), p. 5.
19. Jackson's first geological publication (coauthored by his friend Alger) resulted from their observations in Nova Scotia in 1827 and 1829. An 1833 revised version of this work was referred to by geologist George Featherstonhaugh as "the neatest and best executed work on geology which has been gotten up in the United States." See Merrill (1924), p. 120. However, in 1836 the Canadian geologist Abraham Gesner published, to Canadian acclaim, an almost identically titled work clearly based largely on, but without acknowledgment of, the work of Jackson and Alger. The pair engaged in several attempts to gain the credit which they, naturally enough, felt was due them, but without success. Perhaps this incident, with the sense of having been cheated out of the recognition that he deserved, lodged in Jackson's mind, to be mulled over in the future. Von Bitter (1977, 1978) has discussed this episode and speculated on its relationship to Jackson's later conflicts.
20. *Dictionary of Scientific Biography*, S. V. "Charles Thomas Jackson," by George Edmund Gifford, Jr., p. 45.
21. Ludovici (1961), pp. 61-62. Numbers and Orr (1981) discuss the general reception of Beaumont's work, but do not mention Jackson.
22. Kendall (1852), pp. 15, 46. Ludovici (1961), pp. 63-66, provides the basics of the Morse affair.
23. Bruce (1987), p. 38.
24. Ludovici (1961), pp. 10-12; Raper (1945), p. 156. For a minority report supportive of Jackson's claims and dissenting from a congressional committee's finding in favor of Morton, see Stanly and Evans (1852).
25. Merrill (1924), p. 279.
26. Hubbard to Bradish, July 18, 1850, in Hubbard Papers.
27. J. D. Whitney, in particular, was important in American geology. He studied in Europe for five years and had already had some experience with Jackson on Lake Superior. Jackson was evidently instrumental in steering him away from chemistry and into geology. In 1854 he published *The Metallic Wealth of the United States*, a milestone in the study of ore deposits. In 1860 he assumed the directorship of the California survey, and in 1865 became professor at Harvard. Mt. Whitney is named for him. See Brewster (1909).
28. Jackson (1849b), pp. 403-63, gives his journal account of the survey.
29. Ibid., p. 438.
30. Ibid., p. 716.
31. Merrill (1924), p. 669.
32. Jackson to Walker, July 19, 1848. Photostat in Jackson Papers.
33. Gould to Hall, Mayday, 1849, in Merrill (1924), p. 669.
34. Many of the details of Jackson's conduct on the survey appear in a composite anonymous document that appeared after the messy affair that surrounded his dismissal from the survey, titled *Full Exposure of the*

Conduct of Dr. Charles T. Jackson, Leading to His Discharge from the Government Service, and Justice to Messrs. Foster and Whitney, U. S. Geologists. The bibliography of Hazen and Hazen (1980), p. 219, dates this document to 1849, with no place of publication, and that date is repeated in Hazen and Hazen (1985), pp. 105 and 108. George Edmund Gifford, in *Dictionary of Scientific Biography,* s.v. "Charles Thomas Jackson," p. 45, cites 1851, with publication at Washington. The document is undated, but the Hazen date of 1849 is clearly too early, for dated material written in mid-1850 is included. Unless otherwise noted, the events of the survey and Jackson's fall from it are taken from this work, cited here as *Full Exposure* [1851]. It is interesting to compare this title to that of Kendall's book (previously noted) written at the height of the affair with Morse, *Morse's Patent: Full Exposure of Dr. Chas. T. Jackson's Pretensions* . . . . Fully exposing Jackson evidently required much ink in the mid-nineteenth century.

35. *Full Exposure* [1851], pp. 14–16.
36. Ibid., pp. 7, 14, 15.
37. Ibid., p. 12.
38. Ibid., pp. 8–10.
39. Gould to Hall, Mayday, 1849, in Merrill (1924), pp. 669–71. The settlement of disputes between prominent Bostonians by a select group of their peers was apparently a standard procedure at the time. The previous year Augustus Gould had also been involved in mediating the bitter controversy between Louis Agassiz and his former assistant Edward Desor. See Edward Lurie, *Louis Agassiz: A Life in Science,* (Chicago: University of Chicago Press, 1960), reprint ed., (Baltimore: Johns Hopkins University Press, 1988), p. 156.
40. *Full Exposure* [1851], p. 13.
41. "Official Reply of J. D. Whitney and J. W. Foster to C. T. Jackson's Defence [sic]," in *Full Exposure* [1851], pp. 21–32.
42. *Full Exposure* [1851], pp. 3–7, 17–18.
43. Irving (1883), p. 7.

8. Jackson and the Early Mining Efforts

1. Jackson (1849b). Quote is on p. 719. Others examples of such backbiting are on pp. 708, 710, 715, 726, and 771, the last involving Foster.
2. Sagard was discussed above in chapter 1. Some subsequent writers have continued to refer to a "Lagarde." For examples, see Williams (1907), p. 2; Rickard (1932), p. 222; Drier and DuTemple (1961), p. 37; and Heinrich (1976), p. 11.
3. Jackson (1849b) pp. 385–88.
4. Jackson (1850), pp. 66–67.
5. *Full Exposure* [1851], p. 31.
6. Ibid., pp. 23, 31.
7. Hybels (1950), p. 228.
8. Clarke (1974), p. 4.
9. Jackson (1845a), p. 82.
10. Jackson (1849b), pp. 384–85.

Notes to Chapter 8

11. U.S. Congress, House Doc. 211, 29th Cong., 1st sess., June 17, 1846, pp. 12-13.
12. Jackson (1844), p. 8.
13. Ibid., p. 12.
14. Jackson (1845a), p. 82.
15. Ibid., p. 88.
16. Ibid., pp. 88-89.
17. Jackson (1846a), p. 111.
18. Jackson (1845b), p. 9. It should be noted that in his use of the word *ores* in the last sentence here, Jackson is clearly referring to native copper.
19. Jackson (1845b), p. 8.
20. Ibid., pp. 9-10.
21. Jackson (1844), p. 26.
22. Jackson (1849b), p. 387.
23. Foster and Whitney (1850), pp. 137-38. Jackson did indicate in his report that the rich samples should not be taken as representative of all of the ore. Jackson (1845b), p. 11.
24. Whitney (1854), p. 267.
25. For the story of the Pittsburgh and Boston Company, see Chaput (1971).
26. *American Journal of Science* (1845), 49: 64.
27. Chaput (1971), p. 19.
28. *Niles Register*, May 1845, p. 192, quoted in Hyde (1984), p. 7.
29. Williams (1907), p. 17; Hyde (1984), p. 8.
30. Williams (1907), pp. 13-14.
31. Chaput (1971), p. 22.
32. Williams (1907), p. 16.
33. Hybels (1950), p. 233.
34. Chaput (1971), p. 19, Murdoch (1943), p. 52.
35. Williams (1907), p. 16.
36. Whitney (1854), p. 276.
37. Foster and Whitney (1850), p. 128. In this account, written five years after the events it describes, Whitney also revealed the general commitment of most at that time to the traditional ores: "Up to this period the sandstone and conglomerate were supposed by many to afford the best mining-ground, and that to this source they were to look for permanent supplies of the sulphurets of copper."
38. Ibid. Foster and Whitney called this a "fanciful theory advanced by at least one geologist." That geologist was Jackson, and this seems to be one more example of the sniping between them, noted above, that grew out of their disagreements over the federal survey. Jackson did suggest this theory, but he clearly recognized that the loose boulders of copper found on the surface came originally from the traprock of the district. The only reason he was led to the suggestion that the Ontonagon Boulder was transported from Isle Royale was the report that it had serpentine rock attached to it, a rock that Jackson understood to occur in the region only on the island. Jackson had not seen the boulder, and he also knew that it had been suggested that the attached rock was not serpentine, but he decided to stick with the original claim that it was. It is therefore interesting to note that the claim was made by Houghton himself, who was reflecting the earlier opinion of Schoolcraft.
39. Jackson (1849a), pp. 286-87.

40. Foster and Whitney (1850), p. 139.
41. Whitney (1854), pp. 293-94. Jamison (1950), pp. 69-83 and (1939), pp. 51-54, and Murdoch (1943), chap. 9.
42. Always pronounced "Minnesota" but always written with one *n*. Jamison (1950), p. 69, attributed the spelling to a clerical error when the organizational documents of the mine were drawn up, stating that the owners "were much more interested in making money than they were in preserving the purity of the English orthography." However, the Indian word from which Minnesota comes may well have often been spelled with one *n* in earlier times.
43. Foster and Whitney (1850), p. 135.
44. Foster and Whitney (1850), p. 134.
45. Jamison (1950), p. 69, quoting *Stevens Copper Handbook* (1902), V-2.
46. Murdoch (1943), p. 38.
47. Jackson (1849a), p. 286.
48. Jackson (1849b), p. 610-11.
49. Foster and Whitney (1850), p. 139.
50. *Full Exposure* [1851], p. 3.
51. Foster and Whitney (1850), p. 16.
52. Foster and Whitney (1850). The second part, on the iron lands of the Marquette region, appeared in 1851.
53. Irving (1883), p. 7.
54. Foster and Whitney (1850), p. 180.
55. Ibid., pp. 180-81.

9. The District Becomes Established

1. *Lake Superior Miner*, Feb. 28 and March 7, 1857, as reported in notes of Samuel Brady, superintendent of the Michigan Mine, in the archives of Michigan Technological University, Houghton, Michigan. The details about the size, weight, and date of discovery of this huge mass vary, depending on the source. For some of the variations, see *American Journal of Science* (1864), 37: 431; Stevens (1899), p. 154; Jamison (1939), p. 54, and (1950), p. 80; Murdoch (1943), pp. 90-91; Heinrich (1976), pp. 94-95; and Clarke (1978), No. 11, *Minesota Mining Company*, pp. 19, 22. There is little doubt that its exact weight was never accurately determined.
2. Benedict (1952), pp. 7-10. On the profitability of large masses, see Murdoch (1943), pp. 91-93.
3. Schoolcraft (1855), pp. 292-302. Wording change is on p. 300.
4. Benedict (1952), pp. 7-10.
5. Merk (1982), p. 41.
6. *American Journal of Science* (1841), 41: 183-86. The reference given for this interpretation is an *American Journal of Science* summary of the views that Houghton presented to the Association of American Geologists in 1841, just after the appearance of the copper report.
7. Houghton (1928), pp. 550-51.
8. White (1968), p. 306.
9. Foster and Whitney (1850), p. 167.

10. Whitney (1854), pp. 221, 246. The quote continues, "and associated conglomerates and sandstones," but it is important to recognize that the conglomerates being spoken of are not the conglomerate lodes that were to soon become recognized as being among the most important lodes in the district. The reference here is to conglomerates also cut by the fissures or veins that crossed the trap. These were occasionally mineralized with some copper near the crosscutting vein, but were not generally rich enough to be considered ores, and were not lodes in themselves.
11. Stevens (1899), p. 19.
12. Foster and Whitney (1850), p. 91.
13. Whitney (1854), pp. 264-65.
14. Ibid., p. 297.
15. Stevens (1899), p. 33.
16. The history of the Quincy Mining Company is well told in Lankton and Hyde (1982).
17. Foster and Whitney (1850), pp. 140, 150.
18. Forster (1885), p. 139.
19. Murdoch (1943), p. 115, Stevens (1899), p. 31.
20. Forster (1885), p. 139.
21. Lankton and Hyde (1982), p. 12.
22. Whitney (1854), pp. 301-2.
23. Lankton and Hyde (1982), p. 10.
24. Whitney (1854), p. 302.
25. Lankton and Hyde (1982), p. 21.
26. Benedict (1952), pp. 6-10.
27. The legend is discussed in Benedict (1952), pp. 26-27 and, told with more verve, in Murdoch (1943), chap. 12.
28. Benedict (1952), gives an account of the opening of the Calumet conglomerate in chapter 3. A more dramatic but less documented account is found in Murdoch (1943), chapter 12.
29. Benedict (1952), p. 77, and Murdoch (1943), p. 128.
30. Hulbert called the lode the Calumet-copper-bearing-conglomerate. Well into this century it was known by the Calumet name, and Benedict (1952), in his history of the Calumet and Hecla Company, consistently identifies it as the Calumet conglomerate or Calumet lode. However, in more recent years it has often been referred to as the Calumet and Hecla conglomerate, after the combined companies that worked it so extensively. See "Index of the Nomenclature of Michigan Geological Formations" in Martin and Straight (1956), p. 128.
31. Quoted in Chaput (1971), p. 54.
32. Benedict (1952), chap. 5.

10. The Making of a Mining District

1. Benedict (1952), chap. 6. The story of the Michigan copper-mining industry has recently been masterfully told in Lankton (1991).
2. Wright (1899), p. 131.
3. Murdoch (1943), p. 153.
4. The history of the one major period of labor strife in the Michigan copper district is well told in Thurner (1984).

5. Gates (1951), p. 209, 229.
6. Ibid., p. 163.
7. Heinrich (1976), p. 10.
8. The Quincy steam hoist can be visited by tourists, and the shaft house shown here in figure 29 has recently been resheathed in gleaming new sheet metal. Deciding whether this "upgrade" constitutes an improvement undoubtedly depends on one's philosophy of the past.
9. White (1968), p. 307.
10. Butler and Burbank (1929), pp. 142–46, in their classic paper on the Keweenaw, give a useful summary of other similar deposits known to that date. Brief mention of the Coro Coro deposits can be found in Pennebaker (1954), chap. 2, and a detailed account to 1922 can be found in Singewald and Berry (1922).
11. Benedict (1952), p. 128, and Jamison (1950), p. 59, quoting from *Stevens Copper Handbook* for 1902.
12. Kelly (1984).
13. Native copper, often in broad, thin sheets, is also found as a minor component of the ore body. The mine has had difficulty remaining profitable in recent years, and has changed ownership several times.
14. An imprecise but perhaps revealing illustration of their amateur status and interest in practical results rather than knowledge for its own sake may be seen in the Hazen and Hazen (1980) bibliography of geological literature to 1850. Schoolcraft and Houghton each have 30 entries, while Jackson has 146. There are no entries for Hays, Knapp, Shelden, or Hulbert.

REFERENCES CITED

Abbot, Collamer M. (1971). "Boston Money and Appalachian Copper." *Michigan History* 55: 217-42.
Agassiz, Louis (1850). *Lake Superior: Its Physical Character, Vegetation, and Animals, Compared with Those of Other and Similar Regions.* Boston: Gould, Kendall and Lincoln. Reprint facsimile ed., Huntington, N.Y.: Robert E. Krieger Publishing Co., 1974.
Aldrich, Michele L. (1979). "American State Geological Surveys 1820-1845." In *Two Hundred Years of Geology in America*, pp. 133-43. Edited by C. Schneer. Hanover, N.H.: University Press of New England.
——— (1981). "Charles Thomas Jackson's Geological Surveys in New England, 1836-1844." *Northeastern Geology* 3 (#1): 5-10.
Allen, Roland C. (1915). "Dr. Douglass Houghton." *Michigan Pioneer and Historical Collections* 39: 127-34.
——— (1922). "A Brief History of the Geological and Biological Survey of Michigan: 1837 to 1872." *Michigan History* 6: 675-99.
Benedict, C. Harry (1952). *Red Metal: The Calumet and Hecla Story.* Ann Arbor: University of Michigan Press.
Beveridge, W. I. B. (1950). *The Art of Scientific Investigation.* New York: Random House.
Bornhorst, Theodore J., William I. Rose, Jr., and James B. Paces (1983). *Field Guide to the Geology of the Keweenaw Peninsula, Michigan.* Houghton: Michigan Technological University.
Bowen, Robert, and Ananda Gunatilaka (1977). *Copper: Its Geology and Economics.* London: Applied Science Publishers Ltd.
Bradish, Alvah (1881). "Doctor Douglass Houghton." *Michigan Pioneer and Historical Collections* 4: 97-99.
——— (1889). *Memoir of Douglass Houghton, First State Geologist of Michigan.* Detroit: Raynor and Taylor.
Brebner, John B. (1933). *The Explorers of North America 1492-1806.* Reprint ed., Garden City, N.Y.: Doubleday and Co., 1955.

Bremer, Richard G. (1982). "Henry Rowe Schoolcraft: Explorer in the Mississippi Valley 1818-1832." *Wisconsin Magazine of History* 66: 40-59.

Brewster, Edwin T. (1909). *Life and Letters of Josiah Dwight Whitney*. Boston and New York: Houghton Mifflin Co.

Brooks, T. B. (1873). *Upper Peninsula 1869-1873. Part I I ron-Bearing Rocks (Economic)*. Geological Survey of Michigan. New York: Julius Bien.

Bruce, Robert V. (1987). *The Launching of American Science 1846-1876*. Ithaca, N.Y.: Cornell University Press.

Butler, B. S., and W. S. Burbank (1929). *The Copper Deposits of Michigan*. U.S. Geological Survey Professional Paper 144. Washington: U. S. Government Printing Office.

Carter, Clarence E. (1942-45). *The Territorial Papers of the United States—The Territory of Michigan*. Vols. 10 (1942), 11 (1943), 12 (1945). Washington: U. S. Government Printing Office.

Carter, James L. (1979). "A Trip to Lonely Caribou Island." *Inland Seas* 35:4-12.

Carter, James L., and Ernest H. Rankin, eds. (1970). *North to Lake Superior. The Journal of Charles W. Penney 1840*. Marquette, Mich.: John M. Longyear Research Library.

Cass, Lewis (1824). *Letter from Governor Cass, of Michigan, on the Advantage of Purchasing the Country upon Lake Superior Where Copper Has Been Found*. U. S. Congress, Senate document No. 19, 18th Cong., 2d sess., 1825.

Chaput, Donald (1971). *The Cliff: America's First Great Copper Mine*. Kalamazoo, Mich.: Sequoia Press.

Chase, Lew Allen (1945). "Early Copper Mining in Michigan." *Michigan History* 29: 22-30.

Clarke, Don H. (1973-). *Copper Mines of the Keweenaw*. n.p.

Cleaveland, Parker (1816). *An Elementary Treatise on Mineralogy and Geology*. Boston: Cummings and Hilliard.

—— (1822). *An Elementary Treatise on Mineralogy and Geology. Second Edition*. Boston: Cummings and Hilliard.

Costain, Thomas B. (1954). *The White and the Gold. The French Regime in Canada*. Garden City, N.Y.: Doubleday and Co.

Dana, James D. (1837). *A System of Mineralogy*. New Haven, Conn.: Durrie, Peck, Herrick, and Noyes.

Daniels, George H. (1967). "The Process of Professionalization in American Science: The Emergent Period, 1820-1860." *Isis* 58: 151-66.

—— (1968). *American Science in the Age of Jackson*. New York: Columbia University Press.

Derch, Virginia J. (1977). "Copper Mining in Northern Michigan: A Social History." *Michigan History* 61: 291-321.

Dictionary of Canadian Biography (1966-). 11 vols. Toronto: University of Toronto Press.

Dictionary of Scientific Biography (1970-80), 16 vols. Charles C. Gillispie, ed. New York: Charles Scribner's Sons.

Dorr, John A. Jr., and Donald F. Eschman (1970). *Geology of Michigan*. Ann Arbor: The University of Michigan Press.

Dott, R. H. Jr. (1984). Review of *Earth Sciences History*. *Isis* 75: 217-18.

Drier, Roy W., and Octave J. Du Temple, eds. (1961). *Prehistoric Copper Mining in the Lake Superior Region*. Calumet, Mich.: Drier and Du Temple.

Dunbar, Willis Frederick (1965). *Michigan: A History of the Wolverine State*. Grand Rapids: Eerdmans.

Eaton, Amos (1830). *Geological Text-book, Prepared for Popular Lectures on North American Geology; with Applications to Agriculture and the Arts*. Albany: Websters and Skinners.

——— (1831). "Travelling Term of Rensselaer School, for 1830, with a Notice of the Nature of the Institution." *American Journal of Science* 19: 151–59.

Eccles, W. J. (1972). *France in America*. New York: Harper and Row.

Eustis, William (1818). "Experiments Made by the Assay-Master of the King of the Netherlands, at the Mint of Utrecht, on the Native Copper Existing in Huge Blocks on the South Side of Lake Superior." *American Monthly Magazine and Critical Review* 2: 366–67.

Fadner, Lawrence T. (1966). *Fort Wilkins 1844 and the U. S. Mineral Land Agency 1843*. New York: Vantage Press.

Faul, Henry, and Carol Faul (1983). *It Began With a Stone*. New York: John Wiley and Sons.

Fergusson, William B. (1981). "Eighteenth Century Geologic Studies in Eastern Pennsylvania." *Northeastern Geology* 3: 62–70.

Fisher, Donald W. (1981). "Emmons, Hall, Mather, and Vanuxem—The Four 'Horsemen' of the New York State Geological Survey (1836–1841)." *Northeastern Geology* 3 (#1): 29–46.

Forster, John H. (1885). "Lake Superior Country." *Michigan Pioneer and Historical Collections* 8: 136–45.

Foster, J. W., and J. D. Whitney (1850). *Report on the Geology and Topography of a Portion of the Lake Superior Land District, in the State of Michigan: Part I. Copper Lands*. Washington.

Friedman, Gerald M. (1981). "Geology at Rensselaer Polytechnic Institute: An American Epitome." *Northeastern Geology* 3 (#1): 18–28.

——— (1983). "'Gems' from Rensselaer." *Earth Sciences History* 2 (#2): 97–102.

Full Exposure [1851]. *Full Exposure of the Conduct of Dr. Charles T. Jackson, Leading to His Discharge from the Government Service, and Justice to Messrs. Foster and Whitney, U. S. Geologists*. n.p.

Fuller, George N. (1939). *Michigan—A Centennial History of the State and Its People*, 5 vols. Chicago: Lewis Publishing Co.

Gallatin, Albert (1811). "American Manufactures in 1810." *American Mineralogical Journal* 1:113–20.

Gates, William B. Jr. (1951). *Michigan Copper and Boston Dollars*. New York: Russell and Russell.

Gilpin, Alec R. (1970). *The Territory of Michigan (1805–1837)*. East Lansing: Michigan State University Press.

Green, Mott T. (1985). "History of Geology." *Osiris*, second series 1: 97–116.

Greene, John C. (1984). *American Science in the Age of Jefferson*. Ames: The Iowa State University Press.

Griffin, James B., ed. (1961). *Lake Superior Copper and the Indians: Miscellaneous Studies of Great Lakes Prehistory*. Anthropological Papers, Museum of Anthropology, University of Michigan, No. 17. Ann Arbor: The University of Michigan.

Hazen, Margaret Hindle, and Robert M. Hazen (1985). *Wealth Inexhaustible. A History of America's Mineral Industries to 1850*. New York: Van Nostrand Reinhold Co.

Hazen, Robert M., and Margaret Hindle Hazen (1980). *American Geological Literature, 1669 to 1850.* Stroudsburg, Penn.: Dowden, Hutchinson & Ross, Inc.
Heinrich, E. Wm. (1976). *The Mineralogy of Michigan.* Bulletin 6. Lansing: Geological Survey Division.
Hendrickson, Walter B. (1961). "Nineteenth-Century State Geological Surveys: Early Governmental Support of Science." *Isis* 52: 357-71.
Henry, Alexander (1809). *Travels and Adventures in Canada and the Indian Territories between the Years 1760 and 1776.* New York: I. Riley; reprinted in *March of America Facsimile Series* vol. 43. Ann Arbor: University Microfilms, Inc., 1966.
Houghton, Douglass (1928). *Geological Reports of Douglass Houghton.* Ed. George N. Fuller. Lansing: Michigan Historical Commission.
——— Houghton Papers, Michigan Historical Collections, Bentley Historical Library, University of Michigan, Ann Arbor.
Houghton, Jacob Jr. (1846). *The Mineral Regions of Lake Superior.* Buffalo: O. G. Steele.
Houghton, Jacob Jr., and T. W. Bristol (1846). *Reports of Wm. A. Burt and Bela Hubbard, Esqs. on the Geography, Topography and Geology of the U. S. Surveys of the Mineral Region of the South Shore of Lake Superior, for 1845.* Detroit: Charles Willcox.
Hubbard, Bela (1848). "A Memoir of Dr. Douglass Houghton." *American Journal of Science*—second series 5: 217-27.
——— (1887). *Memorials of a Half-Century.* New York: G. P. Putnam's Sons.
——— Hubbard Papers, Michigan Historical Collections, Bentley Historical Library, University of Michigan, Ann Arbor.
Hybels, Robert James (1950). "The Lake Superior Copper Fever 1841-47." *Michigan History* 34: 97-119, 224-44, 309-26.
Hyde, Charles K. (1984). "From 'Subterranean Lotteries' to Orderly Investment: Michigan Copper and Eastern Dollars 1841-1865." *Mid-America* 66 (January): 3-20.
Irving, Roland D. (1883). *The Copper Bearing Rocks of Lake Superior.* U. S. Geological Survey Monograph 5. Washington.
Jackson, Charles T. (1844). "Report on the Mines and Minerals Belonging to the Lake Superior Copper Company." In *A Brief Account of the Lake Superior Copper Company, by an Original Shareholder,* pp. 8-16. Boston: S. N. Dickinson & Co., 1845.
——— (1845a). "On the Copper and Silver of Keweenaw Point, Lake Superior." *American Journal of Science* 49: 81-93.
——— (1845b). *Dr. Charles T. Jackson's Report to the Trustees of Lake Superior Copper Company.* Boston: Beals and Greene.
——— (1846a). [Jackson on the Lake Superior District, presented March 4, 1846.] *Proceedings of the Boston Society of Natural History* 2: 110-14, 1848.
——— (1846b). "Obituary [of Douglass Houghton]." *American Journal of Science,* second series 1: 150-52.
——— (1849a). "Copper of the Lake Superior Region." *American Journal of Science,* second series, 7: 286-87.
——— (1849b). U.S. Congress, Senate Exec. Doc. 1, *Report on the Geological and Mineralogical Survey of the Mineral Lands of the United States in the State of Michigan,* December 24, 1849. 31st Cong., 1st sess., 3: 371-935. Also House Exec. Doc. 5, 31st Cong., 1st sess., vol. 5.

——— (1850). "On the Geological Structure of Keweenaw Point." *American Journal of Science*, second series 20: 65–77.
——— Jackson Papers, Michigan Historical Collections, Bentley Historical Library, University of Michigan, Ann Arbor.
Jamison, James K. (1939). *This Ontonagon Country*. Ontonagon, Mich.: Ontonagon Herald Co.; reprint ed., Calumet, Mich.: Roy W. Drier, 1965.
——— (1950). *The Mining Ventures of This Ontonagon Country*. Ontonagon, Mich.: Ontonagon Herald Co.
Joralemon, Ira B. (1973). *Copper*. Berkeley, Calif.: Howell-North Books.
Kalm, Peter (1937). *Peter Kalm's Travels in North America*. Trans. Adolph Benson. New York: Wilson Inc.
Karpinski, Louis C. (1931). *Bibliography of the Printed Maps of Michigan*. Lansing: Michigan Historical Commission.
Kellogg, Louise P. (1925). *The French Regime in Wisconsin and the Northwest*. Madison: State Historical Society of Wisconsin.
Kellogg, Louise P., ed. (1917). *Early Narratives of the Northwest 1634–1699*. New York: Scribners Sons.
Kelly, William C. (1984). "Precambrian Oil and Copper of the Upper Peninsula." In *Conference Guidebook, Michigan Earth Science Teachers Association*, 18th annual conference, University of Michigan.
Kendall, Amos (1852). *Morse's Patent—Full Exposure of Dr. Chas. T. Jackson's Pretensions to the Invention of the American Electro-chemical Telegraph*. Washington.
Kohlstedt, Sally Gregory (1976). *The Formation of the American Scientific Community. The American Association for the Advancement of Science 1848–60*. Champaign: University of Illinois Press.
——— (1985). "Institutional History." *Osiris*, second series 1: 17–36.
Kraus, Edward H. (1938). "A Notable Centennial in American Mineralogy." *American Mineralogist* 23: 145–48.
Krause, David J. (1989a). "Testing a Tradition: Douglass Houghton and the Native Copper of Lake Superior." *Isis* 80: 622–39.
——— (1989b). "Henry Rowe Schoolcraft and the Native Copper of the Keweenaw." *Earth Sciences History* 8: 4–13.
Lanctot, Gustave (1963). *A History of Canada*, 3 vols. Cambridge: Harvard University Press.
Lankton, Larry D. (1991). *Cradle to Grave: Life, Work, and Death at the Lake Superior Copper Mines*. New York: Oxford University Press.
Lankton, Larry D., and Charles K. Hyde (1982). *Old Reliable*. Hancock, Mich.: Quincy Mine Hoist Association, Inc.
Laudan, Rachel (1977). "Ideas and Organizations in British Geology: A Case Study in Institutional History." *Isis* 68: 527–38.
Laws of the Territory of Michigan (1835). Detroit: S. McKnight.
Livnat, Alexander (1983). "Metamorphism and Copper Mineralization of the Portage Lake Lava Series, Northern Michigan." Ph.D. dissertation, Department of Geological Sciences, University of Michigan.
Ludovici, L. J. (1961). *The Discovery of Anaesthesia*. New York: Thomas Y. Crowell Co.
Lyon, Lucius (1897). "Letters of Lucius Lyon." *Michigan Pioneer and Historical Collections* 27: 412–604.
McKenney, Thomas L. (1827). *Sketches of a Tour to the Lakes*. Baltimore: F. Lucas, Jr.

Magnaghi, Russell M. (1980). "Aborted Cooper Expedition to Lake Superior." *Inland Seas* 36 (#2): 80–86.
Marshall, Orsamus H. (1887). *Historical Writings of the Late Orsamus H Marshall.* Albany: J. Munsell's Sons.
Martin, Helen M. (1945). "Douglass Houghton (1809–1845)." *Michigan Conservation* 14 (October): 10.
Martin, Helen M., and Muriel Tara Straight, eds. (1956). *An Index of the Geology of Michigan 1823–1955.* Michigan Department of Conservation Publication 50. Lansing.
Martin, John Bartlow. (1944). *Call It North Country.* New York: Alfred A. Knopf. Reprinted 1986, Detroit: Wayne State University Press.
Mather, W. W. (1839). "Report of W. W. Mather of the First Geological District." *American Journal of Science* 36: 15.
Merk, George P. (1981). "The Legacy of Douglass Houghton (1809–1845), Michigan's Pioneer Geologist." *Geological Society of America Program With Abstracts.* Annual Meeting, Cincinnati, Ohio, p. 508–9.
——— (1982). "Douglass Houghton, Michigan's First State Geologist." *Earth Sciences History* 1: 36–44.
Merrill, George P. (1924). *The First One Hundred Years of American Geology* (facsimile of 1924 edition). New York: Hafner Publishing Co., 1964.
Michigan Pioneer and Historical Collections (1874–1929), 40 vols. Lansing: Robert Smith and Co.
Miller, David Philip (1986) "Method and the 'Micropolitics' of Science: The Early Years of the Geological and Astronomical Societies of London." In *The Politics and Rhetoric of Scientific Method*, pp. 227–57. Ed. John A. Schuster and Richard R. Yeo. Dordrecht, Holland: D. Reidel.
Milstein, Randall L. (1987). "Michigan Survey Honors Founder." *Geotimes*, November: 9–10.
Moore, Charles (1895). "The Ontonagon Copper Bowlder in the U. S. National Museum." In *Report of the U. S. National Museum*, pp. 1023–30. Washington: Smithsonian Institution.
Murdoch, Angus (1943). *Boom Copper: The Story of the First U. S. Mining Boom.* New York: Macmillan Co.
Numbers, Ronald L., and William J. Orr, Jr. (1981). "William Beaumont's Reception at Home and Abroad." *Isis* 72: 590–612.
Nute, Grace L. (1944). *Lake Superior.* Indianapolis: Bobbs-Merrill Co.
Osborn, Chase S., and Stellanova Osborn (1942). *Schoolcraft—Longfellow—Hiawatha.* Lancaster, Penn.: The Jaques Cattell Press.
Pennebaker, E. N. (1954). "Copper Minerals, Ores, and Ore Deposits." In *Copper: The Science and Technology of the Metal, Its Alloys and Compounds*, pp. 21–62. Ed. Allison Butts. New York: Reinhold Publishing Corp.
Peters, Bernard C. (1983). *Lake Superior Journal.* Marquette: Northern Michigan University Press.
——— (1985). "The Origin and Meaning of Chippewa and French Place Names along the Shoreline of the Keweenaw Peninsula." *Michigan Academician* 17: 195–211.
Quinn, David B. (1974). *England and the Discovery of America, 1481–1620.* New York: Alfred A. Knopf.
Quinn, David B., ed. (1979). *New American World: A Documentary History of North America to 1612.* 5 vols. New York: Arno Press.

Raper, Howard R. (1945). *Man against Pain*. New York: Prentice-Hall.
Rickard, T. A. (1905). *The Copper Mines of Lake Superior*. New York and London: Engineering and Mining Journal.
——— (1932). *A History of American Mining*. New York and London: McGraw Hill Book Co. Inc.
Rintala, Edsel K. (1954). *Douglass Houghton, Michigan's Pioneer Geologist*. Detroit: Wayne University Press.
Robertson, James M. (1975). "Geology and Mineralogy of Some Copper Sulfide Deposits near Mount Bohemia, Keweenaw County, Michigan." *Economic Geology* 70: 1202–24.
Russell, James, and Albert Hornstein (1899). *First Annual Review of the Copper Mining Industry of Lake Superior*. Marquette, Mich.: Mining Journal Co.
Russell, Nelsen V. (1939). *The British Regime in Michigan and the Old Northwest 1760–1796*. Northfield, Minn.: Carleton College.
Sagendorph, Kent (1947). *Stevens Thomson Mason—Misunderstood Patriot*. New York: E. P. Dutton & Co.
St. John, John R. (1846). *A True Description of the Lake Superior Country*. New York: William H. Graham.
Schoolcraft, Henry Rowe (1819). *A View of the Lead Mines of Missouri Including Some Observations on the Mineralogy, Geology, Geography, Antiquities, Soil, Climate, Population, and Productions of Missouri and Arkansaw, and Other Sections of the Western Country*. New York: Charles Wiley and Co.
——— (1821a). *Narrative Journal of Travels through the Northwestern Regions of the United States Extending from Detroit through the Great Chain of American Lakes to the Sources of the Mississippi River in the Year 1820*. Ed. Mentor L. Williams. East Lansing: Michigan State University Press, 1953.
——— (1821b). "Account of the Native Copper on the Southern Shore of Lake Superior, with Historical Citations and Miscellaneous Remarks, in a Report to the Department of War." *American Journal of Science* 3: 201–16.
——— (1822). *A Report of the Secretary of War, on the Number, Value, & Position, of the Copper Mines on the Southern Shore of Lake Superior*. Senate document No. 5, 17th Cong., 2nd sess., 33 pp.
——— (1824). "Notice of a Recently Discovered Copper Mine on Lake Superior, with Several Other Localities of Minerals." *American Journal of Science* 7: 43–46.
——— (1834). *Schoolcraft's Expedition to Lake Itasca*. Ed. Philip P. Mason. East Lansing: Michigan State University Press, 1958.
——— (1851). *Personal Memoirs of Thirty Years with the Indian Tribes*. Philadelphia: Lippincott, Grambo, and Co.
——— (1855). *Summary Narrative of an Exploratory Expedition to the Sources of the Mississippi River in 1820: Resumed and Completed by the Discovery of Its Origin in Itasca Lake, in 1832*. Philadelphia: Lippincott, Grambo, and Co.
Shepard, Charles U. (1832, 1835). *Treatise on Mineralogy and Treatise on Mineralogy, Second Part*. New Haven, Conn.: Hezekiah Howe and Co., and Herrick and Noyes.
Sherzer, W. H. (1932). "An Unpublished Episode in Early Michigan History." *Michigan History* 16: 214–17.

Singewald, Joseph T., and Edward W. Berry (1922). "The Geology of the Corocoro Copper District of Bolivia." In *Johns Hopkins University Studies in Geology*, no. 1, pp. 1-117. Ed. Edward B. Mathews. Baltimore: Johns Hopkins University Press.

Soule, Annah May (1897). "The Southern and Western Boundaries of Michigan." *Michigan Pioneer and Historical Collections* 27: 346-90.

Stanly, Edward, and Alexander Evans (1852). U.S. Congress, House. *Report to the House of Representatives of the United States of America, Vindicating the Rights of Charles T. Jackson to the Discovery of the Anæsthetic Effects of Ether Vapor, and Disproving the Claims of W. T. G. Morton to That Discovery.* 32d cong., 1st sess. Washington.

Starbuck, James C. (1962). "Ben Franklin and Isle Royale." *Michigan History* 46: 157-66.

Stevens, Horace J. (1899). *First Annual Review of the Copper Mining Industry of Lake Superior.* Marquette, Mich.: Mining Journal Company Limited.

Stoiber, R. E., and E. S. Davidson (1959). "Amygdule Mineral Zoning in the Portage Lake Lava Series, Michigan Copper District." *Economic Geology* 54: 1250-77, 1444-60.

Stuart, L. G. (1897). "Verdict for Michigan: How the Upper Peninsula Became a Part of Michigan." *Michigan Pioneer and Historical Collections* 27: 390-403.

Taylor, John W. (1930). "Reservation and Leasing of the Salines, Lead and Copper Mines of the Public Domain." Ph.D. dissertation, University of Chicago.

Thurner, Arthur W. (1984). *Rebels on the Range: The Michigan Copper Miners' Strike of 1913-1914.* Lake Linden, Mich.: The John Forster Press.

Thwaites, Ruben Gold, ed. (1896-1901). *The Jesuit Relations and Allied Documents*, 73 vols. Cleveland: Burrows Brothers.

——— (1902). *The French Regime in Wisconsin I, 1634-1727.* Madison: Collections of the State Historical Society of Wisconsin, vol. 16.

——— (1906). *The French Regime in Wisconsin II, 1727-1748.* Madison: Collections of the State Historical Society of Wisconsin, vol. 17.

Vogel, Virgil J. (1986). *Indian Names in Michigan.* Ann Arbor: University of Michigan Press.

Von Bitter, Peter H. (1977). "Abraham Gesner (1797-1864), An Early Canadian Geologist—Charges of Plagiarism." *Geoscience Canada* 4: 97-100.

——— (1978). "Charles Jackson, M. D. (1805-1880) and Francis Alger (1807-1863)." *Geoscience Canada* 5: 79-82.

Wallin, Helen (1977). *Douglass Houghton, Michigan's first State Geologist 1837-1845.* Pamphlet 1 revised, Department of Natural Resources. Lansing.

Warner, Robert M. (1970). "The Drowning of Douglass Houghton." *Inland Seas* 26: 118-22.

White, Walter S. (1960). "The Keweenawan Lavas of Lake Superior, An Example of Flood Basalts." *American Journal of Science*, Bradley Volume 258-A, pp. 367-74.

———. (1968). "The Native-Copper Deposits of Northern Michigan." In *Ore Deposits of the United States, 1933-1967*, vol. 1. pp. 303-25. Ed. John D. Ridge. New York: American Institute of Mining, Metallurigical, and Petroleum Engineers, Inc.

References Cited

Whitney, Josiah Dwight (1854). *The Metallic Wealth of the United States, Described and Compared with That of Other Countries.* Philadelphia: Lippincott, Grambo & Co.

Whittlesey, Charles (1863). "Ancient Mining on the Shores of Lake Superior." In *Smithsonian Contributions to Knowledge,* V-13, No. 155: 1–29. Washington, D.C.

Williams, Ralph D. (1907). *The Honorable Peter White.* Cleveland: Penton Publishing Co.

Winchell, Alexander (1886?). "Michigan. First Geological Survey under Douglass Houghton." In *Contributions to a History of American State Geological and Natural History Surveys,* U. S. National Museum Bulletin No.109, pp. 158–203. Ed. George P. Merrill. Washington: Smithsonian Institution, 1920.

——— (1889). "Douglass Houghton." *American Geologist* 4: 128–39.

Wold, Richard J., and William J. Hinze, eds. (1982). *Geology and Tectonics of the Lake Superior Basin.* Geological Society of America Memoir 156.

Woodworth, J. B. (1897). "Charles Thomas Jackson." *American Geologist* 20: 78–110.

Wright, J. N. (1899). "The Development of the Copper Industry of Northern Michigan." *Publications of the Michigan Political Science Association* 3: 127–41.

Wrong, George M., ed. (1939). *The Long Journey to the Country of the Hurons, by Father Gabriel Sagard.* Publications of the Champlain Society, vol. 25. Toronto.

Young, Otis E., Jr. (1984). "The American Copper Frontier, 1640–1893." *The Speculator: A Journal of Butte and Southwest Montana History* 1 (Summer): 4–15.

INDEX

Acacia, 24, 25
Adams, John, 61, 62
Adams, John Quincy, 92
Agassiz, Louis, 270n.39
Alger, Francis, 182, 269n.19
Algonquin Indians, 24, 25, 26
Allen, James, 108–9, 110, 111
Allouez (priest), 30
American Academy of Arts and Sciences, 192–93
American Association for the Advancement of Science, 55, 123, 196
American Fur Company, 127, 129
American Indians, 14, 164, 208, 252n.1; Cass and, 65, 67–68, 71, 91, 95; colonial hostilities with, 23, 24, 32, 33–34, 38, 39; copper mines, 24, 25, 27, 30, 34, 38, 52, 214–15, 228, 232, 234, 247; copper uses, 20, 24, 195, 253n.4; early explorers and, 20, 21–22, 23, 24, 25, 26, 30, 39; Expedition of 1820 and, 67, 71, 73, 74, 76, 77; Expedition of 1831 and, 103, 104; Expedition of 1832 and, 108, 109; Houghton and, 109, 126–27; Lake Superior population, 122; land cessions, 135; land titles, 61–62, 91–92, 93, 112; mineral rights cession, 93, 94; missionaries and, 28; sale of Ontonagon Boulder, 136; Schoolcraft and, 69–70, 73, 76, 85, 95, 96; Treaty of Fond du Lac, 92, 93, 125; Treaty of La Pointe, 135; Treaty of Prairie du Chien, 92; tribal wars, 24, 25, 28, 95, 96
American Journal of Science, 100, 256n.10; Association of Geologists accounts, 157; on copper-mining potential, 159; Expedition of 1820 report, 79, 83; Houghton obituary in, 149, 197–98; on Houghton's geological survey, 119, 135, 150, 158; Hubbard's memoir of Houghton, 153; Jackson's articles in, 196, 197–98, 201; and professionalization of science, 56, 79
American Revolution, 41, 61
Amygdaloid lodes, 255–56n.2; copper production from, 47, 48, 233, 236, 239, 244, 250–51; formation of, 45, 227; Houghton on, 161, 225, 233; Jackson on, 200–201, 205; mining of, 48, 226, 230, 232–33, 237, 239, 242
Ann Arbor, Mich., 120, 141
Apostle Islands, 31, 122
Arkansas, 70

285

Association of American Geologists and Naturalists, 123, 135, 142, 156–59
Atlantic seaboard, 20
Atwatanik, 29
Austria, 159

"Barrel-work," 217, 221
Basalt, 44, 225
Baxter, Alexander, Jr., 39, 40
Beaumont, William, 184–85
Beck, Lewis, 99
Benton, Thomas Hart, 90–91, 92
Beveridge, W. I. B., 173
Black River, 34
Bolivia, 244–46, 247
Boom Copper (Murdoch), 11
Bostwick, Henry, 39
Botany, 104, 108
Boucher, Pierre, 29–30
Boudrie, John Baptiste, 143, 144, 146, 147
Bradish, Alvah: on Houghton's attitude toward science, 172–73, 263n.5; idealized depictions of Houghton, 144–45, 146, 260n.15; on "opposition" to Houghton's geology, 154, 156–57, 158, 160; portraits of Houghton, 98
Brazil, 159
Bressani, Francesco, 28–29
Bristol, T. W., 177
British Museum, 40, 256n.6
Brockway, "Dad," 207, 209
Brulé, Etienne, 26, 27, 29
Buffalo, Chief, 126
Burt, William A., 142, 143, 176–77, 181, 188, 194
Butterfield, Justin, 219

Cabot, John, 19
Cabot, Sebastian, 19
Calhoun, John C.: and Expedition of 1820, 67, 68, 70, 71; Schoolcraft's reports to, 77, 79, 82, 83, 85, 88, 89, 223; as secretary of war, 66–67, 68
California gold rush, 15
Calumet Company, 236, 237
Calumet conglomerate, 236, 237, 243, 244, 273n.30

Calumet and Hecla Mining Company, 237, 239, 240, 242–43
Canada, 41, 42
Carbonate ores, 48, 251; Houghton on, 165, 170, 171; Jackson on, 201; Schoolcraft on, 80, 81, 82, 84
Caribou Island, 40
Carter, Nathaniel, 80, 82
Cartier, Jacques, 21–22, 23, 24
Carver, Jonathan, 41, 42
Cass, Lewis, *illus.* 66, 88; in Expedition of 1820, 71, 72, 74, 76, 77; as governor of Michigan, 65–66, 163; on Houghton's death, 150; and Indian land titles, 91, 92, 93; and Ontonagon Boulder, 74, 77, 92, 103, 138; organization of Expedition of 1820, 65, 67–68, 71; as secretary of war, 105, 107, 108; sympathy for Indians, 65, 67, 71, 91, 95; and Upper Peninsula issue, 114–15
Catholic priests, 26, 127. See also Jesuits
Central mine, 217, 224
Chalcocite, 247
Chalcopyrite, 48
Champlain, Samuel de, 24, 25, 26, 28
Chegoimegon, Point, 77
China, 29
Chippewa Indians, 38, 74, 92, 93, 95
Chocolate (Chocolay) River, 116, 124–25, 133
Christianity, 28
Chrysocolla, 207
Civil War, 233
Clarke Historical Library, 128
Cleaveland, Parker, 83, 101, 159
Clemens (colonel), 94
Cliff Mine: copper production of, 210, 212, 217, 218, 224, 225, 226, 237; discovery of, 209–10, 244; dividend payments, 222, 233; mining at, 172, 187, *illus.* 213, 216
Columbus, Christopher, 16, 18, 19, 175
Conglomerate lodes, 255–56n.2; Calumet conglomerate, 236, 237, 243, 244, 273n.30; copper production from, 47, 48, 234, 236, 250–51; formation of, 45, 50,

Index

51, 227; Foster and Whitney on, 228, 272n.10; Houghton on, 106, 133, 135, 161, 162, 225, 233; Jackson on, 201; mining in, 48, 226, 233-34, 235-36, 271n.37
Connecticut, 195
Cook, George, 99
Cooper, Richard Fenimore, 62
Copper: amateur and professional scientists and, 16-17, 55, 56, 57-58, 59, 84, 87, 198-99, 251; amygdaloid lode content, 230, 232, 233, 244; British production of, 48-49, 88; chemical properties, 48; conglomerate lode content, 233-34, 236, 244; European analysis of, 64, 77-78, 159; European explorers and, 18-20, 23, 25-26, 27-28, 29-31, 33, 34-38, 39-41, 58, 247-48; Expedition of 1820 and, 65, 67-68, 71, 73, 76; Expedition of 1831 and, 103, 104, 105-7, 111; Foster and Whitney report on, 219-20, 226, 227, 228, 230; geological formation of, 44, 50-52, 226, 246-47; in "green rock" vein, 88-89, 104, 111, 126, 207; in Houghton's copper report, 134, 135, 156, 165, 166, 171, 225; Houghton's geological view of, 149, 151-52, 160-70, 171-72, 173, 174-75, 196, 204, 224-25, 249; Houghton's ideas on, "opposition" to, 152-53, 154-58, 203-4; Indian descriptions of, 21-22, 23, 24-25, 30, 38; Indian mineral-right cession, 93, 94; Indian use of, 20, 30, 52, 195, 247; industrial uses of, 61; Jackson's claim of discovery, 193, 195-97, 198, 199-202, 204-5, 249-50, 251; lode types, 45-48, 224-25, 255-56n.2; Michigan production of, 15, 212, 227, 239, 240, 241, 242; mineralogy texts and, 159-60; native state, uniqueness of Keweenaw, 16, 17, 49-50, 51, 55, 84, 152, 220, 244, 246, 247, 251; of Nova Scotia, 182; in Ontonagon Boulder, 74-75, 78, 92-93, 95, 110, 128; ore types, 48-49; prices, 233, 240; Schoolcraft reports on, 77-78, 80, 81-84, 85, 86-88, 90, 159, 170, 223-24, 248; secondary production, 243; in sedimentary rocks, 246-47; sizes of masses mined, *illus.* 53, 212, 214, 216, 221-22, *illus.* 223; South American deposits, 199, 244-46, 247; state geological survey of, 118, 125, 129; U.S. government and, 60, 61-63, 91, 92, 93, 94, 114; U.S. imports of, 15, 49, 61, 64; U.S. production of, 15, 49, 240, 242; vein structures, 226-28; western U.S. deposits, 241, 244, 255-56n.2. *See also* Mining; Ores
Copper Falls, 228
Copper Harbor, Mich., 187; black copper oxide ore, 208, 210, 211, 218; copper-rush arrivals at, 143; establishment of, 122; federal mineral agency at, 137; "green rock" vein, 88-89, 125, 163, 171, 207, 208
"Copper report" (Houghton, 1840), 133-34, 140, 156, 165-68, 225; contribution to copper rush, 132, 134-35, 163
Copper rush, 15, 143, 154, 181, 211; Houghton's copper report and, 132, 134-35, 163; Jackson's survey reports and, 197; lease-permit system and, 140
Corbin (carpenter), 35
Cordilleras mines, 199
Cornwall mines, 170; copper ores, 48-49, 165, 206, 226, 227; copper production, 49, 88; Houghton's comparisons with, 164, 165, 167, 172
Coro Coro, Bolivia, 244-46, 247
Cradle to Grave (Lankton), 15-16
Cuba, 202
Cunningham, Walter, 137, 171

Dablon (priest), 30-31, 32
Dana, James, 56, 157, 159, 160

De la Ronde, Louis Denis, Sieur, 34, 35-37, 38, 40, 255n.41
Desor, Edward, 270n.39
Detroit, Mich., 77; ceded to United States, 43; Houghton and, 102, 104-5, 141; University of Michigan moved from, 120; and Upper Peninsula issue, 112, 114, 115
Detroit Daily Advertiser, 135
Detroit Free Press, 115
Diamonds, 18-19, 23, 24
Digestion, 184-85
Dollier (priest), 32
Douglas, Theresa, 230
Douglass, Columbus Christopher; in Houghton's geological survey, 124, 127, 132; in Jackson's explorations, 200, 201; mining efforts, 230, 231-32, 233
Douglass, David, 71, 74, 76
Douglass, Mary Lydia, 97
Druillettes (priest), 29
"Dykes," 135, 162

Eagle River, 129, 147, 203, 205, 209-10, 211
Earth Sciences History, 161
Eaton, Amos, 57, 99, 100-101, 102, 117, 165, 172
Eldred, Julius, 136-38, 263-64n.31
Electric industry, 61
Emerson, Ralph Waldo, 182
Emmons, Ebenezer, 99, 101
Erie, Lake, 23, 113
Erosion, 47, 81
Ether anesthetic, 184, 186, 190
Ethnology, 70
Europe, 182; copper mining in, 89, 153, 154, 166, 202, 206; geological surveys, 142; interest in Keweenaw copper in, 64, 85, 159; landholding in, 179-80; "savants of," 153, 174
European explorers, 19, 27-28; interest in copper, 14, 52, 58, 247-48
Eustis, William, 62, 63, 64
Ewing, Thomas, 190, 191, 219
Expedition of 1820, 65, 68, 70, 71-77, 88, 95, 170
Expedition of 1831, 103-4, 163
Expedition of 1832, 108-11
Expedition of 1840, 255n.48

Featherstonhaugh, George, 269n.19
Felch, Alpheus, 177
Fissure lodes, 255-56n.2; copper production from 48, 226, 233, 244, 250-51; formation of, 47-48, 227; Houghton on, 224-25; mining in, 224-26, 233, 237
Fond du Lac, 92, 93, 94, 109
Fond du Lac, Treaty of (1826), 92, 93, 94, 112, 125, 134
Forster, John Adam, 36, 37
Foster, John W.: conflict with Jackson, 189-90, 191-93, 194, 196-97, 204, 205, 206-7, 271n.38; copper report of 1850, 150-51, 193, 213, 219-20, 226, 227, 228, 229, 230; federal geological survey, 187, 188, 189, 192, 193, 216, 218-19; on geology as science, 197, 251; on Houghton, 150-51
France: American colonization, 24, 28; American exploration, 18, 20, 21, 23-24, 26; cession of colonies to Britain, 38-39; copper mining efforts, 32, 34-35, 52-53, 195, 254n.31
Francis I (king of France), 20, 21, 24
Franklin, Benjamin, 42-43
Fundy, Bay of, 25
Fur trade, 29, 67

Gage, Thomas, 39
Galinée (priest), 32
Gallatin, Albert, 64
Geological formation, 44-48, 50-51
Geological Society of London, 256-57n.14
Geological surveys: combined linear-geological survey, 142-143, 176, 178-79, 181-82; federal geological survey under Jackson, 182, 186-92, 193, 194; federal linear survey, 141-42; Foster and Whitney report, 193, 219-20, 226, 229; Jackson report, 193, 194-97; of Michigan, 118-21,

Index   289

148, 178; Michigan "copper report" of 1840, 122, 132–35, 140, 156, 163, 165–68, 225; Michigan, field season of 1840, 124–32, 163; Michigan, final report, 121, 130, 141, 149, 150–51, 176, 178, 183, 265n.6; Michigan, Houghton's annual reports, 119, 121, 122, 130, 140, 152, 197; New York survey, 124, 130; state surveys, 117–18, 130, 150, 183

Geological Textbook (Eaton), 101

Geology, 54, 251; amateur observation, 55, 57; Houghton and, 99, 106, 123–24, 133, 140, 144, 150, 153–54, 158, 160–61, 173, 174, 249; Jackson and, 182–83, 197, 249; organizations, 123; professionalization of, 17, 55, 56, 57, 123; Rensselaer School and, 99, 101; Schoolcraft and, 70, 79, 94; textbooks, 101

Georgian Bay, 23

Geotimes, 161

Germany, 159

Gesner, Abraham, 269n.19

Glacial deposits, 133

Glaciers, 229

Gold, 15, 18–20, 21–22, 23, 24

Gomes, Estavo, 20

Gould, Augustus A., 190, 191, 270n.39

Grand Marais, Mich., 124

Grand Sable Banks, 72, 124

Granite, 106, 110

Grassin (assayist), 35

Gratiot, Charles H., 200, 206

Gray, Andrew, 198–99

Great Britain: American colonies, 41, 43; American exploration, 18, 19; capture of Quebec, 38–39; colonial copper mining efforts, 39–40; colonial fur trade, 67; copper export restrictions, 61; copper mines in, 48–49, 166, 199; French colonial cession to, 39; Indian allies, 71; War of 1812, 63, 65

Great Depression, 242

Great Lakes, 14, 20–21, *map* 21, 146, 254n.26; American control of, 61; copper discovery and, 60, 61; European exploration, 23; Expedition of 1820, 67

Green Bay, Wis., 77

"Green rock" vein ("la roche verte"), 31, 73, 109; composition of, 89, 104, 111, 163, 207; Houghton and, 125, 126, 171, 201, 207, 210; Jackson on, 201; mining of, 201, 207, 208–9; Schoolcraft and, 88–89, 90, 104, 106, 207

Greenstone lava flow, 44

Greenstone Ridge, 211

Grenoble (voyageur), 26, 27

Groseilliers (fur trader), 29

Guncotton, 184

Hall, James, 99, 118, 147, 157

Hammond (colonel), 136, 137

Hays, John, 207, *illus.* 208, 209, 210–11, 251

Hecla Company, 236, 237

Henry, Alexander, 39, 40, 106, 254n.31

Henshaw, David, 197, 199–200

Higgins, S. W., 132, 177, 178, 188

Hill, Samuel W., 190

Hitchcock, Edward, 117–18, 123, 158

Hochelaga, 21, 22, 24

Houghton, Douglass, 12, 17, 54, *illus.* 98, 229, 230; and amateur and professional science, 57, 59, 99, 101–2, 123, 155, 172–73, 174, 248–49; *American Journal* obituary, 149, 197–98; and amygdaloid and conglomerate lodes, 106, 133, 161, 162, 225, 233, 250; in Association of American Geologists, 123–24, 135, 140–41, 142, 156–58; association with Keweenaw copper, 97, 107, 140, 175, 266–67n.46; Bradish's *Memoir* of, 144–46, 154, 156–57, 172–73, 260n.15; and combined linear-geological survey, 141–44, 178, 181, 186; on copper ores, 105–6, 107, 121, 132, 135, 162, 163–65, 166,

167, 169–71, 172, 173, 210, 213, 267n.55, 268n.68; death of, 144, 146–48, 149–51, 174–75, 176–77; early life, 97–99; elected mayor of Detroit, 141; in Expedition of 1831, 96, 103–5; Expedition of 1831 report, 105–7, 115; in Expedition of 1832, 108–11; geological training, 57, 99–101, 248; geological view of native copper, 149, 151–52, 160–70, 171–72, 173, 174–75, 196, 198, 204, 211, 224–25, 249, 267nn.55 and 57; geological views, "opposition" to, 152–53, 154–58, 159, 203–4; and the "green rock," 125, 126, 171, 201, 207, 210; "Houghton tradition," 152, 153, 156, 159, 171–72, 174, 175, 198, 224–25; Jackson and work of, 188, 189, 195, 196–98, 202, 251; marriage, 112; medical profession, 111–12; Michigan, relocation to, 102, 157; as Michigan state geologist, 11, 112–13, 118, 119, 140, 210; and Michigan statehood, 114, 116–17, 261–62n.59; and Ontonagon Boulder, 104, 106, 107, 110, 127, 128, 138, 163, 167, 271n.38; real estate investments, 112, 263n.5; state geological survey, 118–21, 124–32, 148, 158; state survey, annual reports, 119, 121, 122, 130, 140, 152, 197; state survey, "copper report" of 1840, 122, 132–35, 140, 156, 165–68, 225; state survey, final report, 121, 130, 141, 149, 150–51, 176, 178, 183, 265n.6; on traprock formation, 105–7, 133, 162, 163, 260n.27; at University of Michigan, 120, 141

Houghton, Harriet Stevens, 112, 144
Houghton, Jacob, 97
Houghton, Jacob, Jr., 177
Houghton County, Mich., 226
Hubbard, Bela, 123, 181, 263n.5; contribution to copper report, 132; in geological survey of Michigan, 124, 125, 127, 145; on Houghton, 153, 160; and Jackson, 186, 187, 188, 189; and survey final report, 176–77, 178, 194
Hubbard, Fredrick, 124, 132
Hudson, Henry, 24
Hudson Bay, 24
Hughes, G. W., 170
Hulbert, Edwin J., 234–37, *illus.* 235, 243, 251
Huron, Lake, 23, 26, 28, 34, 71
Huronia, 28
Huron Indians, 28, 29

Ice Age, 51, 133, 229
Illinois, 103
Iron deposits, 28–29, 48, 142, 263n.7
Iron oxide, 202
Iron pyrite, 19, 24
Iron River, 30, 31, 35, 255n.39
Iroquois, Point, 104
Iroquois Indians, 24, 28, 29
Isle Royale (Minong), 188, 212, 271n.38; copper deposits, 52; copper mining on, 27, 30, 31, 147; geological structure, 45, 46, 133; Houghton's exploration of, 129, 133; Indians on, 27, 29, 30; U.S. boundary and, 42–43
Isle Royale Mine, 232

Jackson, Andrew, 105
Jackson, Charles T., 11, 17, 54, *illus.* 183, 224, 229, 269n.27; and amateur and professional science, 59, 183–84; appointed U.S. Geologist, 182, 187; attacks on Foster and Whitney, 192–93; claims of priority in discoveries, 184–86, 190, 196, 249; discovery claim on native copper, 193, 195–97, 198, 199–202, 204–5, 249–50, 251; education, 182; federal survey of Michigan, 187–89; federal survey report, 192, 193, 194–95, 197; federal survey resignation, 190–91, 219; Foster's and Whitney's charges against, 189–90, 191–92, 196–

Index 291

97, 205, 206-7; geological publications, 182-83, 199, 201, 269n.19; on the "green rock," 201; an insane institution, 186; involvement in mining operations, 201, 206, 209, 212, 249; on mining potential for copper, 195, 196, 200-202, 204, 205, 206, 211, 212, 218, 250; Nova Scotia and New Jersey geological studies, 182-83, 249, 253n.4; obituary of Houghton, 197-98; on the Ontonagon Boulder, 201, 271n.38; opposition in Michigan to, 182, 186-87, 188; as state geologist of Maine, Rhode Island, and New Hampshire, 183; use of Houghton's work, 151, 188, 189, 197, 198
Jacobsville sandstone, 45
Jefferson, Thomas, 62
Jesuit *Relations*, 28, 30, 31, 33
Jesuits, 28-29, 30-31, 38; map of Lake Superior, 31-32, 33, 254n.26
Johnson, Sir William, 39
Johnston, Jane, 69-70
Jolliet, Louis, 32
Joy, Charles, 190

Kalm, Peter, 38
Keweenaw Bay, 73
Keweenaw copper district, 11, map 200; amateur and professional science and, 16-17, 54, 55, 59, 84, 248, 250-51; chalcocite ore mining, 247; copper lodes, 45-48, 51, 225-26, 237-38, 250, 255-56n.2; copper ore formation, 48, 49, 50; copper producton, 15, 49, 221-22, 225, 227, 238, 240-41, 242, 244; early mining efforts, 200, 205-9, 211-12, 224, 229, 248, 250; European exploration of, 14-15, 16, 29, 64, 247-48; Expedition of 1820 and, 71, 73, 77, 88; Foster and Whitney on, 220; geological formation of, 44-45, 46, 47, 50-51, 161, 162, 226-27, 228; glacial formations, 51, 229; Houghton and, 12, 17, 97, 102, 107, 140, 155, 160-62, 167-70, 172, 174-75, 203-4, 225; Indian land titles, 90-91; Indians of, 14, 20, 247; Jackson and, 17, 193, 194-95, 198, 202, 204-5, 249-50; Michigan acquisition of, 15, 112-13, 114, 116; Michigan geological survey of, 121, 122, 133; mining boom, 15, 138, 211, 221; native copper, understanding of significance of, 155, 159, 160, 167-70, 172, 174-75, 198-99, 202-3, 248; native copper, uniqueness of, 16, 17, 43, 49, 203-5, 220, 221, 244, 246, 247, 251; role in U.S. development, 12, 15, 88; Schoolcraft and, 17, 71, 84, 88, 90, 96; U.S. acquisition of, 43, 60; U.S. government interest in, 60, 61, 90
Keweenaw County, 226
Keweenaw fault, 45
Keweenaw peninsula, 31, 207; in Carver's map, 41; geological formation of, 44, 45, 46, 47, 107, 133, 182; Indians and, 14; Michigan geological survey of, 143; portage route, 29, 88
Keweenaw Point, 147, 226; copper mines, 105, 195, 201, 227; copper ores of, 88, 125, 163; Expedition of 1831 and, 104, 105, 106; "green rock" vein, 73, 88-89, 104, 109, 125, 201; Jackson's article on, 201; Michigan geological survey on, 125; vein structures, 227, 230
Knapp, Samuel O., 214, 215-16, 232, 251

Lac la Belle, Mich., 212-13, 218-19
Lahontan, Louis-Armand de Lom d'Arce, Baron de, 32
Lake Superior Copper Company, 197, 200, 203, 205-6, 207-8, 209, 212; stamping mill, 201, *illus.* 204, 206

Land sales, 65–66, 181, 182, 192, 219
Lankton, Larry, 15–16
La Pointe, Wis., 122, 127
La Pointe, Treaty of (1843), 135
La Tourette (fur trader), 32
Lava flows: copper lodes, 47, 161, 162, 225; formation of, 44, 45, 50, 51
Lead mining, 29, 70, 138, 180
Lead ore, 180
Lease-permit system, 138–40, 179–80
Le Baron, Francis, 62–63, 64
Lederer, Baron, 141
Le Sueur (trader), 32
Locke, John, 187, 188
Lodes, 45–48, 225, 237–38, 255–56n.2
Longfellow, Henry Wadsworth, 70
Louis XV (king of France), 34
Lower Peninsula, 42, 112, 120, 127, 141–42
Lyell, Charles, 123
Lyon, Lucius, 108, 114–15, 116, 176–77, 181, 188

McFarland, Peter, 143, 144, 146–47
McKenney, Thomas L., 92–93, 94
Mackinac, Straits of, 113, 114
Mackinac fort. *See* Michilimackinac fort
Mackinac Island, 71
Madeline Island, 127
Madison, James, 65
Maine, 183
Malachite, 89, 163
Manitou Island, 255n.48
Manual of Mineralogy and Geology (Emmons), 101
Marquette, Mich., 72–73, 125
Martin, Helen M., 156, 157
Mason, Stevens T., 116, 118
Mather, William, 118
Memoir of Douglass Houghton (Bradish), 144
Metallic Wealth of the United States (Whitney), 226, 269n.27
Miami (Maumee) River, 113
Michigan: Cass and development of, 65–66, 67; copper district acquisition, 61, 112–13, 114–17; copper production, 240, 241, 242; copper resources, 114, 115, 159, 165; federal geological survey, 181, 186; Houghton and, 97, 102, 124, 148, 150, 151–52, 249; and mineral land sales, 180, 182; mining boom, 15, 197; population growth, 118; state geological survey, 118–21, 130, 148, 149, 158, 176, 178; statehood, 61, 112–15, 116; Territory, establishment of, 61, 113; Upper Peninsula acquisition, 113, 114, 115, 261–62n.59; Wisconsin border dispute, 129, 140
Michigan, Lake, 28, 77, 103, 113
Michigan legislature, 119–20, 177–78, 180
Michilimackinac fort, 39, 43, 62, 63, 67
Mineralogy: Houghton and, 152, 153, 154, 156; Jackson and, 202; Schoolcraft and, 68–69, 70, 81, 82, 83, 85, 88, 90, 104; textbooks, 83, 101, 159–60
Mineral rights, 93, 94
Minesota Mining Company, *illus.* 218, 271–72n.42; copper production, 216–17, 221–22, 224, 225; dividend payments, 222, 233, 237
Minesota vein, 214–16, 218, 227, 232
Mining: in amygdaloid lodes, 48, 226, 230, 232–33, 237, 239, 242; at Central mine, 217, 224; at Cliff mine, 210, 211–12, 218, 222, 225, 226, 237; in conglomerate lodes, 48, 226, 233–34, 236, 237, 271n.37; copper boom, 15, 134–35, 140, 181, 197, 206; of copper ores, 84, 88, 206, 209, 211, 213, 218, 226, 240; copper production, 15, 49, 209, 212, 217, 222, 225, 227, 239, 240–41, 242, 244; demise of, 15–16, 237, 243–44; early European efforts, 32, 34, 36–38, 39, 40, 43, 106, 248; federal survey of, 187; in fissure lodes, 48, 224–26, 233, 237; Foster and

Whitney on, 213, 216, 218-20, 224, 226, 229; "green rock" vein, 201, 207, 208-9; Houghton and development of, 122-23, 134-35; Houghton on potential for, 106, 149, 150, 153, 155, 167-68, 170, 172, 198, 268n.68; by Indians, 52, 214-15, 228, 232, 233, 247; investments in, 177, 181, 197, 206, 222, 224, 239-40; Jackson and development of, 197, 200, 205-6; Jackson on potential for, 195, 196-97, 201, 204-5, 211, 250; in Keweenaw Point, 226, 227; lease-permit system, 138-40, 179-80; Michigan production, 15, 212, 227, 239, 240, 241, 242; at Minesota mine, 216-17, 218, 221-22, 224, 225, 237; native copper, 16, 50, 167, 170, 172, 174-75, 203, 210, 211, 217, 222, 224; native copper as negative indicator for, 166, 167, 198, 206; in Ontonagon region, 226, 227; in Portage Lake district, 226, 227, 229-30, 232, 233; profitability, 212, 216, 217, 222, 233, 237, 239, 242; Schoolcraft on potential for, 78, 84, 85, 86-88, 90, 170, 223-24, 258n.50; U.S. government and, 94, 170, 179; U.S. production, 15, 49, 240, 242; western U.S., 241, 244, 255-56n.2

Mining companies, 187, 224, 231, 233, 236, 237, 239-40, 243

Mining Gazette, 237

Missipicouatong (Michipicoten) Island, 30

Mississippi River: lead mining, 138, 180; source of, Schoolcraft and discovery of, 69, 76-77, 95, 107, 108, 109, 110, 111

Missouri, 70

Mitchell, Samuel, 64

Monroe, James, 65-66, 70, 85

Monroe County, Mich., 115

Montreal (Cascade) River, 116, 125, 129

Mont Royal (Montreal), 22, 24

Morse, Samuel F. B., 185

Morton, William T. G., 186, 192

Murdoch, Angus, 11

Narrative Journal of Travels (Schoolcraft), 72, 75

National Mine, 228

New France, 24, 28, 32, 33

New Hampshire, 183

New Jersey, 135, 195, 249

New York (state), 130

New York, N.Y., 20, 87

Niagara Falls, 24

Nonesuch mine, 246

Norburg (explorer), 40

Norse explorers, 19

Northwest Ordinance (1787), 60, 113

Northwest Territory, 60-61, 62, 159

Norvell, John, 114

Nott (Union College president), 100

Nova Scotia, 182-83, 195, 249

Ohio, 113, 114, 115, 116, 261-62n.59

"On the Copper and Silver of Kewenaw Point" (Jackson), 201

Ontario, Lake, 23, 24

Ontonagon, Mich., 122

Ontonagon Boulder, 59, *illus.* 139, 159-60, 171, 211; attempted removal of, 92-93, 94, 136; Eldred's purchase and removal of, 136-38, 214, 263-64n.31; European explorers and, 25-26, 27, 30, 31, 35, 37, 39, 40; Expedition of 1820 and, 67, 71, 72, 74-75, 76, 77, 78; Expedition of 1831 and, 104; geological origin of, 52, 107, 163, 212, 271n.38; Houghton and, 104, 106, 107, 110, 126, 127, 128, 138, 163, 167, 271n.38; Indians and, 67, 93, 127, 136; Jackson on, 201, 271n.38; mining attempts near, 40, 95, 106, 109-10, 214; Schoolcraft and, 74-75, 78, 92, 94, 136, 137, 159, 160, 271n.38; in Smithsonian Institution, 138, 139; weight estimates of, 74-75, 92-93, 110, 138

Ontonagon County, 226

Ontonagon Indians, 93

Ontonagon River: copper deposits on, 34, 36, 37, 39, 62, 78, 89, 107; copper mining on, 35, 40, 105, 214, 226, 227; European exploration of, 31, 35, 39, 40; Expedition of 1820 to, 71, 73–74, 77; land titles on, 91, 92, 93; transportation obstacles, 87, 116

Ores: British mines, 48–49, 164, 165, 199, 206, 226; copper content, 49, 90, 164, 167, 206, 209; copper production from, 89, 209, 221, 225, 237, 240, 241, 242, 244; Foster and Whitney on, 218, 219–20, 226–27, 271n.37; geological formation, 50, 195, 201–2; "green rock" vein, 88–89, 104, 111, 163, 171, 207, 210; Houghton and, 105–6, 107, 121, 132, 135, 162, 163–65, 166, 167, 169–71, 172, 173, 210, 213, 267n.55, 268n.68; Jackson on, 195, 201–2, 206, 209; mining of, 84, 88, 206, 209, 211, 213, 218, 226, 240; native copper as negative indicator of, 166, 167, 170, 198, 206; native copper as "token" of, 82, 83, 89, 202; processing waste, 243; professional geologists and, 83, 84, 88, 195, 199; Schoolcraft on, 81, 82, 83, 84, 88, 89–90, 170; smelting, 20, 49, 219–20; terminological usage, 162–64, 165, 255–56n.2; types of, 48–49, 256n.7; uniqueness of Keweenaw, 43, 50, 199, 220. See also Carbonate ores; Oxide ores; Sulfide ores

Ottawa River, 22–23
Owen, David Dale, 117–18
Oxide ores, 48, 83; in Cornwall mines, 206; at the "green rock," 104, 111, 171, 208, 210; Houghton on, 165, 169–70, 171; mining of, 208–9, 210, 211, 218, 251

Paleontology, 118
Paris, Treaty of (1763), 39
Paris, Treaty of (1783), 41–42
Paul, James K., 263–64n.31
Penny, Charles W., 124, 125, 127–29, 145, 163–64
Peré, Jean, 32
Pewabic amygdaloid, 233
Philadelphia Exposition (1876), 53
Pictured Rocks, 72, 124, 145
Pigs, 234
Pittsburg and Boston Mining Company, 207–8, 209, 210
Plover, Chief, 93
Polk, James K., 181
Porcupine Mountains, 110, 121, 246
Porcupines, 129
Portage Lake, 73, 88, 226, 243–44; copper production, 227, 229, 237, 239, 242; Houghton's map of, 129–30, 131; Shelden and mining development at, 230–32, 233
Portage Lake Volcanics, 44, 45, 133
Portage River, 73
Porter, Augustus, 134, 164
Porter, George F., 94
Porter, James, 137
Prairie du Chien, Treaty of (1826), 92
Preston, William, 114
Prévert, Sieur de, 25
Protestant missions, 127
Pumpelly, Ralph, 234

Quartz, 19, 24
Quebec, 38–39
Quincy Mining Company, 229–30, 231, 232; copper production, 233, 239, 242, 243; decline of, 243–44, illus. 245, 273n.8

Radisson (fur trader), 29
Railroads, 178
Raymond (speculator), 207, 209, 210
Recollect monks, 26, 28
Rensselaer County, N.Y., 117
Rensselaer School, 56, 99–100, 101, 165, 172
"Report on the Existence of Copper in the Geological Basin of Lake Superior" (Houghton), 105–6
Report on the Geology and Topography of a portion of the Lake Su-

Index   295

perior Land District (Foster and Whitney), 219-20
Rhode Island, 183
Rogers, Robert, 39, 41
Rogers, William, 123
Roxbury Smelting Works, 209
Royal, Billy, 234

Sagard, Gabriel, 26, 27, 195
Saginaw Indians, 71
Saguenay kingdom, 22-23
Saguenay River, 21-23
St. Anne River, 37, 255n.39
St. John, John R., 152-53, 156, 160, 203-4
St. Lawrence, Gulf of, 21
St. Lawrence River, 20-21, map 21, 22, 25
St. Louis River, 76
St. Marys River, 169
Salt, 118, 119
Sandstone, 45, 106, 133, 236; copper deposits in, 135, 201-2, 228, 246, 271n.37; Pictured Rocks, 72, 124
Sandy Lake, 76
Sault Ste. Marie, 26, 28, 71, 105, 112, 113, 122, 171
Schönbein, C. F., 184
Schoolcraft, Henry Rowe, 11, 17, 54, illus. 69, 229, 234, 257n.19; and amateur and professional science, 59, 69, 70, 79-80, 83-84, 85, 90, 183, 248, 258n.44; on copper-mining potential, 78, 84, 85, 86-88, 90, 170, 223-24, 248, 250, 258n.50; education, 68, 70, 81, 85, 248; in Expedition of 1820, 70-71, 72, 73, 74-75, 76-77, 95, 170; Expedition of 1820 report, 77-79, 81, 82-85, 90-91, 159, 223-24, 258n.50; in Expedition of 1831, 95-96, 103, 104, 106; in Expedition of 1832, 108, 109, 110, 111; and the "green rock," 89, 90, 104, 106, 207; as Indian agent, 69, 85, 95; and Indian land titles, 85-86, 92; interest in Indian studies, 69-70, 73, 76; interest in mineralogy, 68-69, 70, 104; and Michigan's Upper Peninsula acquisition, 114-115; and Mississippi River source, 69, 76-77, 95, 107, 108, 109, 110, 111; Missouri lead-mining study, 70, 71; on native copper and ores, 80-81, 82-84, 85, 86-87, 88, 89-90, 107, 159, 170, 223-24, 248; and Ontonagon Boulder, 74-75, 78, 92, 94, 136, 137, 159, 160, 271n.38; and Treaty of Fond du Lac, 92, 93; and University of Michigan, 136, 141
Shale, 236, 246
Shelden, Ransom, 230-32, illus. 231, 233, 251
Shepard, Charles, 159-60
Silliman, Benjamin, 56, 79, 83, 123; supposed opposition to Houghton, 153, 157, 158-59
Silver: in copper ores, 40, 89, 134, 200-201, 211, 240; European explorers and, 19-20, 40; Jackson on, 195, 200-201, 206
Sioux Indians, 38, 95
Smallpox, 109
Smelting, 16, 20, 29, 49, 50, 88, 219-20
Smithsonian Institution, 138, 139
"Song of Hiawatha" (Longfellow), 70
South America, 199, 244
Spain, 18, 19, 20, 24
Stamping equipment, 201, illus. 204, 233
"Stamp sand," 243
Stamp-work, 221
Stevens, Harriet, 112
Stockton, John, 171
Sulfide ores, 48; copper content, 49, 241, 255-56n.2; formation of, 50-51, 246-47; Houghton on, 165; Jackson on, 202; mining of, 50, 212-13, 218-29, 220, 240, 241, 244, 246, 251, 252n.4; Schoolcraft on, 80, 81, 82, 83, 84; smelting, 49, 219-20, 247
Sulfur, 48, 50-51, 246-47
Summary Narrative (Schoolcraft), 79, 223
Superior, Lake, 14, 23, map 27, map

33, map 42, map 72; canal proposal, 105, 169; copper deposits, 17, 25-26, 30-31, 44, 49, 64, 111, 154-55, 170, 199, 207, 227, 228, 247; copper-mining industry, 15-16, 39, 43, 149, 196, 217-18, 221, 222; copper rush, 15, 62, 135; European explorers and, 20, 26, 27-28, 29, 30, 32, 34, 35, 36, 37, 39, 41; Expedition of 1820 and, 65, 67, 71, 73, 76, 88; Expedition of 1831 and, 103; federal land survey on, 182, 186; fisheries, 115; Foster and Whitney report on, 207, 213, 219, 220; geological formation, 44, 45, 47, 51, 52, 107; Houghton's survey on, 129, 132, 134, 141, 144, 198; Indian inhabitants, 28, 29, 122, 247; Indian land titles, 61, 85, 90, 91; Jackson's report on, 201; Jesuit map of, 31-32, 33, 254n.26; lease-permit system, 180; St. John's travels on, 152; Sault Ste. Marie rapids and, 105, 112; Schoolcraft report on, 78, 79, 80, 87
Swansea, Wales, 49, 219-20
Swedish copper, 64, 85
Syncline, 45

Talon, Jean, 31, 32
Tecumseh, 65
Telegraph, 184, 185
Toledo War, 113, 114, 261-62n.59
Torrey, John, 108, 109, 111, 120, 157
Tracy (senator), 67
Transportation, 32, 39, 40, 87, 112, 229, 246
Traprock, 110; copper ores in, 105-6, 107, 163, 182, 195, 205, 212, 271n.38; formation of, 44, 162; Houghton on, 105-7, 133, 162, 163, 260n.27; Jackson on, 195, 201-2, 205; mining in, 217, 228, 229; Schoolcraft on, 81
Travels through the Interior Parts of North America (Carver), 41

Treatise on Mineralogy (Cleaveland), 83
Trowbridge, Charles, 73, 76
True Description of the Lake Superior Country (St. John), 152

United States: copper consumption, 61, 87; copper imports, 15, 49, 61, 105; copper land-title acquisition, 61-62, 71, 91, 135; copper mineral-rights acquisition, 93, 94; copper production, 15, 97, 239, 240, 242; geological surveys, 117, 179; independence, 60, 61; mineral land sales, 182; northern boundary settlement, 41-43, 60; Ontonagon Boulder claim, 137, 138
United States Congress, 105; House Committee on Public Lands, 180; House report on Houghton, 149-50; and Indian land titles, 61-62, 85; Jackson survey report, 194, 199, 219; and lease-permit system, 180; and Michigan boundaries, 114, 115; and Michigan copper lands, 61, 85, 86, 170, 171; Michigan delegation, 186, 188; and mineral land sales, 181, 182; Ontonagon Boulder act, 138; Senate, 79, 85, 86, 114, 170, 171; Senate Committee on Roads and Canals, 169
United States War Department, 68, 77-78, 108, 138
Universal Oil Products, 243
University of Leyden, 64, 77-78
University of Michigan, 120, 136, 141
Upper Peninsula, 15, 207; Expedition of 1820 in, 72-73; Houghton's copper report on, 132, 133; iron deposits, 263n.7; Michigan statehood controversy over, 114-16, 118, 261n.45, 261-62n.59; Michigan Territory boundaries in, 113; Michigan-Wisconsin boundary, 140; state geological survey in, 119, 120,

142, 148, 164, 165, 176, 178; Treaty of Paris and, 41-42

Van Rensselaer, Stephen, 99, 117
Verdigris, 31, 33
Verrazzano, Giovanni da, 20
Victoria Mining Company, 258n.35
View of the Lead Mines of Missouri (Schoolcraft), 70

Walker, Robert J., 182, 188, 189
Ward, George L., 190
War of 1812, 65
Wells, Horace, 186
Western U.S., 241, 244, 255-56n.2
White Pine mine, 247, 252n.4
Whitney, Josiah D., 271n.37; conflict with Jackson, 189-90, 191-93, 194, 196-97, 204, 205, 206-7, 271n.38; copper report of 1850, 150-51, 193, 213, 219-20, 226, 227, 228, 229, 230; federal geological survey, 187, 188, 189, 192, 193, 211, 216, 218-19; on geology as science, 197, 251; on Houghton, 150-51; *Metallic Wealth of the United States*, 226-27, 232, 269n.27
Whittlesey, Charles, 215
Wilkins, William, 170
Wilkins, Fort, 208
Winchell, Alexander, 150, 178, 268n.68
Wisconsin, 103, 112, 114, 116, 129, 140
Woodbridge (governor), 129, 132, 133, 138, 164
Woolsey, Melancthon, 103

Yellow Sands, Island of, 40, 255n.48

www.ingramcontent.com/pod-product-compliance
Lightning Source LLC
Chambersburg PA
CBHW070724160426
43192CB00009B/1308